125 Advances in Polymer Science

Springer-Verlag Berlin Heidelberg GmbH

Statistical Mechanics Deformation Ultrasonic Spectroscopy

With contributions by
R. B. Bird, S. V. Bronnikov, C. F. Curtiss, S. Y. Frenkel,
N. Hiramatsu, K. Matsushige, H. Okabe, V. I. Vettegren

With 66 Figures and 7 Tables

Springer

ISBN 978-3-662-14783-2 ISBN 978-3-540-47673-3 (eBook)
DOI 10.1007/978-3-540-47673-3

Originally published by Springer-Verlag Berlin Heidelberg New York in 1996.
Softcover reprint of the hardcover 1st edition 1996
Library of Congress Catalog Card Number 61-642

Typesetting: Macmillan India Ltd., Bangalore-25
SPIN: 10508848 02/3020 - 5 4 3 2 1 0 - Printed on acid-free paper

Editors

Table of Contents

Statistical Mechanics of Transport Phenomena: Polymeric Liquid Mixtures

C.F. Curtiss
Theoretical Chemistry Institute, Department of Chemistry,
University of Wisconsin-Madison, Madison, WI 53706, USA

R. Byron Bird
Department of Chemical Engineering and Rheology Research Center,
University of Wisconsin-Madison, Madison, WI 53706, USA

A summary is given of the kinetic theory of flexible macromolecules, represented by bead-spring models of arbitrary connectivity. The formal theory is applicable to polydisperse systems, multicomponent mixtures, dilute or concentrated solutions, and fluids with concentration, temperature, and/or velocity gradients. Formal expressions are given for the momentum-flux (stress) tensor, the mass-flux vector, and the heat-flux vector; these can be combined with stochastic simulations to solve problems in rheology, diffusion, and heat conduction. Care is taken to point out where empiricisms are introduced, so that these can be modified or eliminated in the future. Other topics included are: the equation of change for angular momentum, the elastic terms in the equation of change for energy, the effect of velocity gradients on thermal conduction, the uniqueness of the molecular expression for the stress tensor, the relation between the mass flux and the stress tensor, thermal diffusion, and the Onsager receiprocal relations.

Advances in Polymer Science, Vol. 125

List of Symbols

(Symbols frequently used and equation number where first introduced; " + " or " – " indicates that symbol occurs in text after or before equation)

A_{ij}	Elements of the Rouse matrix	2.12
$\mathbf{a}$	$\nabla \ln T$	12.3
$B_{\nu k}, \bar{B}_{k\nu}$	Matrix elements	2.2, 2.4, 2.6, 2.7
$B(x)$	Function in phase space	3.2
$\mathbf{b}_\alpha$	$\nabla \ln n_\alpha$	12.4
$b_\nu^{\alpha i}$	Function in momentum space	5.3
C_v	Heat capacity	C.17
C_{ij}	Elements of the Kramers matrix	2.12
$C_\nu^{\alpha i}$	Function in configuration space	5.3
$C_\nu^{\alpha i, \beta j}$	Function in configuration space	5.12
$\bar{C}_\nu^{\alpha}$	Average value	5.3
$\bar{C}_\nu^{\alpha, \beta}$	Average value	5.12
c_j	Eigenvalue of the Kramers matrix	2.12 +
D_{tr}	Translational diffusivity	14.23 + , 15.7
D_{ijk}	Matrix elements	13.3 +
D/Dt	Substantial derivative	13.6 +
$\mathbf{E}$	Electric field intensity	14.28 –
$\mathbf{F}_\nu^{(\phi)\alpha i}$	Intramolecular force on bead	2.28
$\mathbf{F}_\nu^{(e)\alpha i}$	External force on bead	2.29
$\mathbf{F}_\nu^{(\Phi)\alpha i}$	Intra- and intermolecular force on bead	2.30
$\mathbf{F}_\nu^{(d)\alpha i}$	Intermolecular force on bead	2.31
$\mathbf{F}_\nu^{(b)\alpha}$	Brownian force on bead	11.5, 12.16, 12.17
$\mathbf{F}_\nu^{(h)\alpha}$	Hydrodynamic force on bead	11.6, 12.6
$\mathbf{F}_\nu^{\alpha i}$	Total force on bead	2.32
$\mathbf{F}^{(e)\alpha i}$	Total external force on center of mass	2.34
$f(x, t)$	Distribution function in phase space	3.1
f_α	Distribution function in singlet phase space	4.1
$f_{\alpha\beta}$	Distribution function in doublet phase space	4.6
$\mathbf{G}$	External force per unit volume	7.1
$\mathbf{G}^\alpha$	External force per unit volume for species	15.6
H	Hookean spring constant	2.1 –
J	Work done by diffusional effects	8.1
$\mathbf{j}_\alpha$	Mass-flux vector for species	6.1
k	Boltzmann's constant	12.8
m_ν, m_ν^α	Bead mass	2.1, 2.14 +
m_m, m_m^α	Molecule mass	2.1, 2.14 +
N, N_α	Number of beads in molecular model	2.1 – , 2.14 +
n_α	Number density of centers of mass of species	4.2
$\mathbf{p}_\nu^\alpha, \mathbf{p}_\nu^{\alpha i}$	Momentum of bead	2.5

$\mathbf{p}^{\alpha}, \mathbf{p}^{\alpha i}$	Sets of momenta	2.16, 2.19
$\mathbf{Q}$	Interbead vector in dumbbell model	2.1 −
$\mathbf{Q}_k^{\alpha}$	Relative position vector	2.2
$\mathbf{Q}^{\alpha}, \mathbf{Q}^{\alpha i}$	Sets of relative position vectors	2.17
$\mathbf{Q}_k'^{\alpha}$	Normal coordinates in chain models	13.8, 13.11
$Q^{(k)}, Q^{(\Phi)}, Q^{(e)}$	Source terms in energy equation	8.7, 8.8, 8.9
$\mathbf{q}$	Heat-flux vector	8.1
$\mathbf{q}_{\alpha}$	Species contribution to heat-flux vector	16.26
$\mathbf{q}^{(k)}$	Kinetic contribution to heat-flux vector	8.11
$\mathbf{q}^{(\phi)}$	Intramolecular contribution to heat-flux vector	8.18
$\mathbf{q}^{(e)}$	External force contribution to heat-flux vector	8.14
$\mathbf{q}^{(d)}$	Intermolecular contribution to heat-flux vector	8.20
q	Electric charge	14.28 −
$\mathbf{R}_{\nu}^{\alpha}, \mathbf{R}_{\nu}^{\alpha i}$	Position vectors relative to center of mass	2.4, 2.14 +
$\mathbf{R}_{\nu\mu}^{\alpha i}, \mathbf{R}_{\nu\mu}^{\alpha i, \beta j}$	Interbead vectors	2.20, 2.21
$\mathbf{R}_{\alpha\beta}$	Intermolecular distances	2.22
$\mathbf{r}$	Location in three-dimensional space	4.2
$\mathbf{r}_{\nu}^{\alpha}, \mathbf{r}_{\nu}^{\alpha i}$	Bead position vectors	2.1, 2.14 +
$\dot{\mathbf{r}}_{\nu}^{\alpha}, \dot{\mathbf{r}}_{\nu}^{\alpha i}$	Bead velocity vectors	2.5 −
$\mathbf{r}_{c}^{\alpha}, \mathbf{r}_{c}^{\alpha i}$	Molecule center-of-mass position vectors	2.1, 2.14 +
$\mathbf{r}_{c}^{\alpha\beta}$	Location of center of mass of molecule pair	2.23
$\mathbf{r}^{\alpha}, \mathbf{r}^{\alpha i}$	Sets of position vectors	2.15, 2.18
$\mathbf{S}, \mathbf{S}^{(\phi)}, \mathbf{S}^{(d)}, \mathbf{S}^{(e)}$	Momentum sources	7.4
$\mathbf{T}$	Torque	9.1
T	Absolute temperature	12.3
T_{ν}^{α}	Temperature at bead location	12.8
t	Time	3.1
U	Internal energy per unit volume	C.2
$\hat{U}$	Internal energy per unit mass	8.1, C.1
$U_{\nu}^{\alpha i}$	Internal energy associated with bead	8.2
$u_{\alpha, eq}$	Contribution to equilibrium internal energy	C.15 +
$\mathbf{u}_{\nu}^{\alpha}$	Averaged bead velocity	11.2
$\mathbf{v}(\mathbf{r}, t)$	Mass-average fluid velocity	6.3
$\mathbf{v}_{\alpha}$	Species velocity	6.1
x	Set of phase-space coordinates	3.1
$\mathbf{Y}_j$	Contributions to relative velocity vector	16.4
$\boldsymbol{\alpha}$	Tensor in dumbbell distribution function	13.5
$\boldsymbol{\alpha}_j$	Tensors in Rouse distribution function	13.8
$\boldsymbol{\alpha}^{(0)}, \boldsymbol{\alpha}^{(1)}$	Contributions to the α tensor	13.21
$\mathbf{B}$	Finger tensor	13.7
$\boldsymbol{\beta}_{\alpha}$	Tensor in heat-flux vector expression	16.34
$\boldsymbol{\Gamma}; \boldsymbol{\Gamma}_j$	$\boldsymbol{\delta} - \boldsymbol{\alpha}; \boldsymbol{\delta} - \boldsymbol{\alpha}_j$	13.7, 13.10
$\dot{\boldsymbol{\gamma}}$	Rate-of-deformation tensor	14.19, 16.25
$\dot{\gamma}$	Shear rate (a scalar)	13.22
$\boldsymbol{\gamma}_{[0]}$	Relative finite strain tensor	13.7

$\boldsymbol{\Delta}_\alpha$	Diffusivity tensor	15.10
$\delta(x), \delta(\mathbf{r})$	Dirac delta function	4.1
δ_{ij}	Kronecker delta	2.6
$\boldsymbol{\delta}$	Unit tensor with components δ_{ij}	12.18
$\dot{\varepsilon}$	Elongation rate (a scalar)	13.24, 14.21
ζ, ζ_ν^α	Bead friction factor	12.6
η	Viscosity	14.20
η_s	Solvent viscosity	14.20
$\bar{\eta}$	Elongational viscosity	14.21
$\boldsymbol{\kappa}$	Transpose of the velocity-gradient tensor	12.2
λ	Thermal conductivity	16.33
λ_H	Time constant for Hookean dumbbell model	13.6
λ_j	Time constants for Rouse chain model	13.10
λ_s	Solvent contribution to thermal conductivity	16.33
$\boldsymbol{\Xi}$	Tensor virial multiplied by 2	A.7
Ξ	Momentum space distribution function	12.11
ξ	Integration variable in Taylor series	5.11, 5.21
$\boldsymbol{\pi}$	Stress tensor (momentum flux tensor)	7.1
$\boldsymbol{\pi}^{(e)}$	External force contribution to stress tensor	7.12
$\boldsymbol{\pi}^{(k)}$	Kinetic contribution to stress tensor	7.8
$\boldsymbol{\pi}^{(\phi)}$	Intramolecular contribution to stress tensor	7.15
$\boldsymbol{\pi}^{(d)}$	Intermolecular contribution to stress tensor	7.17
ρ	Fluid density	6.2
ρ_α	Species mass concentration	6.1
$\boldsymbol{\sigma}_\alpha$	Tensor in heat-flux vector expression	16.34
$\boldsymbol{\tau}_\alpha$	Species contribution to extra stress tensor	14.13
Φ	Potential energy for all molecules in liquid	2.24 –
$\boldsymbol{\Phi}$	Tensor used in heat-flux expression	16.26
$\phi^{\alpha i}$	Potential energy for single molecule	2.24 –
$\phi^{(e)\alpha i}$	Potential energy for single molecule in external field	2.24 –
$\phi_{\nu\mu}^{\alpha i}, \phi_{\nu\mu}^{\alpha i, \beta j}, \phi_{\nu\mu}^{(d)\alpha i, \beta j}$	Interbead potential energies	2.24, 2.26, 2.27
$\Psi_\alpha, \bar{\Psi}_\alpha$	Singlet configurational distribution functions	4.4, 4.3
$\Psi_{\alpha\beta}, \bar{\Psi}_{\alpha\beta}, \tilde{\Psi}_{\alpha\beta}$	Pair configurational distribution functions	4.8, 4.7, 4.12
ψ_α	Singlet configurational distribution function	13.5, 13.8
$\boldsymbol{\Omega}$	Tensor in relative velocity vector	16.4
Ω_{ij}	Chain-space rotation matrices	13.11

Special Notation

$[[\ ]]^\alpha, [[\ ]]^{\alpha\beta}$	Momentum space averages	5.1, 5.2
$\mathbf{I\alpha\alpha I}_{mnpq}$	Fourth-order tensor with components $\alpha_{mn}\alpha_{pq} + \alpha_{mp}\alpha_{nq} + \alpha_{mq}\alpha_{np}$	13.32, 14.18
$\mathbf{A}^\dagger$	Transpose of the tensor $\mathbf{A}$	8.1a

$\mathbf{A}_{(1)}$	Convected derivative of tensor **A**	13.6
∇, $\partial/\partial\mathbf{r}$	Del operator in 3-dimensional space	6.1, 6.8 +
$\partial/\partial\mathbf{r}_\nu$	Del operator based on bead coordinates	2.13
$\partial/\partial\mathbf{Q}_k$	Del operator based on relative position vector components	2.13
(1), (2), (3)	First-, second-, third-order terms in Taylor expansions	14.4 − , 15.1

Indices

$\alpha, \beta, \gamma \cdots$	Labeling for chemical species	2.14 +
$i, j, k \cdots$	Labeling for enumerating molecules	2.14 +
$\nu, \mu, \eta \cdots$	Labeling for beads in model	2.1
s	Subscript label for solvent	14.20
eq	Subscript label for equilibrium quantity	A.6, C.8
g	Subscript label for gradient contributions	C.11

1 Introduction

In 1944 Kramers [1] published a phase-space kinetic theory for the steady-state potential flow of monodisperse dilute polymer systems in which the polymer molecule is modeled as a freely jointed bead-rod chain. Subsequent scholars developed kinetic theories for shearing flows of monodisperse dilute polymer solutions: Kirkwood [2] for freely rotating bead-rod chains with equilibrium-averaged hydrodynamic interaction, Rouse [3] and Zimm [4] for freely jointed bead-spring chains, and others. These theories were all formulated in the configuration space of a single polymer chain.

In 1975 Curtiss, Bird, and Hassager [5] developed a more general kinetic theory formulated in the phase space of the entire polymeric liquid, the theory being patterned after the Irving and Kirkwood [6] theory for monatomic liquids. The theory includes the possibility of more than one polymer species, polydispersity, models with or without internal constraints, and both dilute and concentrated systems. The Curtiss-Bird-Hassager theory was summarized later in textbook chapters [7, 8] and also applied to polymer melts, both monodisperse [9] and polydisperse [10].

The general theory is useful because:

a. It includes earlier theories as special cases and clarifies the relations among them.
b. It makes clear what assumptions and approximations have been made in the earlier theories.
c. It provides a common framework for studying rheological, diffusional, and thermal phenomena.
d. It provides molecular theory expressions for the fluxes that can be used in conjunction with Brownian dynamics and molecular dynamics simulations.
e. It provides the starting point for the study of solvent-polymer interactions, polydispersity effects, nonhomogeneous systems, cross effects, wall effects, structure-property relations, and other topics of current interest.

The general phase-space theory is regarded as difficult to understand, because it was first presented [1, 5] for mechanical models that may have internal constraints (constant bond lengths and/or constant bond angles). The attendant notational and mathematical complexity tends to obscure some of the physical ideas.

In this article we review the theory of transport phenomena in polymeric liquids in the framework of the phase-space kinetic theory for models with no internal constraints. That is, we restrict ourselves to "bead-spring models," but we do allow for any kind of connectivity; hence the treatment here allows for chains, rings, stars, combs, and branched chains. This presentation includes formally both mixtures and polydisperse systems. We have expanded the coverage here to include the heat-flux vector, which was not investigated in [5], [7], or [8]. Insofar as possible, we use the same notation as that used in the second edition of Dynamics of Polymeric Liquids [8], hereinafter referred to as "DPL." For a broader literature review of polymer transport phenomena and other topics from a variety of viewpoints, see the recent review article by Bird and Öttinger [10d].

Even when constraints are not incorporated in the molecular models, the notation is unavoidably complicated. Many symbols will carry three indices: one for the number of the "bead" on the molecular model (ν, μ, η, ...); one for the chemical species of the molecule(α, β, γ, ...); and one for the number of the molecule of that particular chemical species (i, j, k, ...) in the liquid. Symbols involving the interactions of pairs of beads will involve six indices, three for each bead. Although the notation is unwieldy, it is unambiguous.

Sects. 2–5 deal primarily with notation and definitions that are needed throughout the remainder of the presentation. The main physical result is the "general equation of change" presented in Sect. 3. Special cases of this general equation are then featured in the next five sections, in Sects. 6–9, where the four main conservation equations of fluid dynamics are obtained, with expressions for the fluxes as "by-products," and in Sect. 10, where the equation for the singlet distribution function is developed. The latter is then used in Sect. 11 to obtain the equation of motion for the beads, in which expressions emerge for the Brownian and hydrodynamic forces. Up to this point the development is relatively free from assumptions.

In Sect. 12 five assumptions are introduced in order to carry the development further. In Sect. 13 the equation for the singlet distribution is solved for simple models, both for isothermal and nonisothermal conditions; the solutions are given in terms of the Finger strain tensor, which describes the kinematics of the fluid motion.

Then in the next three sections we continue the discussions given in Sects. 6–8 to show how the general flux formulas can be applied to simple molecular models. In Sect. 14 we obtain the stress-tensor expression for the Rouse model; for the Hookean dumbbell model we obtain the constitutive equation for nonisothermal situations. Also the stress tensor for electrically charged Rouse chains is discussed briefly. In Sect. 15 the diffusivity is found for arbitrary bead-spring models, and the diffusivity tensor in flowing media is found for the Rouse chain model; it is seen how the Soret (thermal diffusion) effect is predicted from kinetic theory. In Sect. 16 it is shown how to get the thermal conductivity in nonflow systems including the Dufour (diffusion-thermo) effect, and also the thermal conductivity tensor for flowing systems; specific results are then given for steady shear and steady elongational flows.

In Sect. 17, we summarize some of the conclusions and speculate on the ways in which the material presented here may be used in the future.

In Appendix A we address some problems that arise in connection with the uniqueness of the expression for the stress tensor. In Appendix B we derive a fairly general stress-diffusion relation for polymer solutions. Finally Appendix C deals with an equation of change for the temperature.

The formal results for the mass, momentum, and energy fluxes to lowest order are summarized in Table 1; the complete expressions are given in Table 2. These results are then used to obtain specific expressions for the fluxes in terms of the driving forces (concentration, velocity, and temperature gradients) for several simple molecular models. The equation numbers giving the locations of these results are as follows:

	Concentration Gradients	Velocity Gradients	Temperature Gradients
Mass Flux	(15.6)	(15.16), (B.17)	(15.6)
Momentum Flux	(14.26)	(14.12)	(14.27)
Energy Flux	(16.26), (16.31)	(16.26)	(16.26), (16.31)

Clearly there is much to be done by using more realistic models, by doing Brownian dynamics simulations, by considering polydispersity effects, and ex-

Table 1. Lowest-order flux contributions for polymeric liquids

Kinetic contributions:

(6.11) $$\mathbf{j}_\alpha = \sum_{\nu} m_\nu^\alpha \int [[\dot{\mathbf{r}}_\nu^\alpha - \mathbf{v}]]^\alpha \Psi_\alpha d\mathbf{Q}^\alpha$$

(7.8) $$\boldsymbol{\pi}^{(k)} = \sum_{\alpha\nu} m_\nu^\alpha \int [[(\dot{\mathbf{r}}_\nu^\alpha - \mathbf{v})(\dot{\mathbf{r}}_\nu^\alpha - \mathbf{v})]]^\alpha \Psi_\alpha d\mathbf{Q}^\alpha$$

(8.11) $$\mathbf{q}^{(k)} = \sum_{\alpha\nu} \tfrac{1}{2} m_\nu^\alpha \int [[(\dot{\mathbf{r}}_\nu^\alpha - \mathbf{v})\cdot(\dot{\mathbf{r}}_\nu^\alpha - \mathbf{v})(\dot{\mathbf{r}}_\nu^\alpha - \mathbf{v})]]^\alpha \Psi_\alpha d\mathbf{Q}^\alpha + \tfrac{1}{2} \sum_{\alpha\nu\mu} \int \phi_{\nu\mu}^\alpha [[\dot{\mathbf{r}}_\nu^\alpha - \mathbf{v}]]^\alpha \Psi_\alpha d\mathbf{Q}^\alpha + \tfrac{1}{2} \sum_{\alpha\beta\nu\mu} \iiint \phi_{\nu\mu}^{(d)} [[\dot{\mathbf{r}}_\nu^\alpha - \mathbf{v}]]^{\alpha\beta} \tilde{\Psi}_{\alpha\beta} d\mathbf{R}_{\alpha\beta} d\mathbf{Q}^\alpha d\mathbf{Q}^\beta$$

External force contributions:

(7.12) $$\boldsymbol{\pi}^{(e)} = \sum_{\alpha\nu} \int \mathbf{R}_\nu^\alpha \mathbf{F}_\nu^{(e)\alpha} \Psi_\alpha \mathrm{d}\mathbf{Q}^\alpha$$

(8.14) $$\mathbf{q}^{(e)} = \sum_{\alpha\nu} \int \mathbf{R}_\nu^\alpha \mathbf{F}_\nu^{(e)\alpha} \cdot [[\dot{\mathbf{r}}_\nu^\alpha - \mathbf{v}]]^\alpha \Psi_\alpha d\mathbf{Q}^\alpha$$

Intramolecular force contributions:

(7.15) $$\boldsymbol{\pi}^{(\phi)} = \tfrac{1}{2} \sum_{\alpha\nu\mu} \int \mathbf{R}_{\mu\nu}^\alpha \mathbf{F}_{\nu\mu}^{(\phi)\alpha} \Psi_\alpha d\mathbf{Q}^\alpha$$

(8.18) $$\mathbf{q}^{(\phi)} = \tfrac{1}{2} \sum_{\alpha\nu\mu} \int \mathbf{R}_{\mu\nu}^\alpha \mathbf{F}_{\nu\mu}^{(\phi)\alpha} \cdot [[\dot{\mathbf{r}}_\nu^\alpha - \mathbf{v}]]^\alpha \Psi_\alpha d\mathbf{Q}^\alpha$$

Intermolecular force contributions:

(7.19) $$\boldsymbol{\pi}^{(d)} = \tfrac{1}{2} \sum_{\alpha\beta\nu\mu} \iiint \mathbf{R}_{\mu\nu}^{\beta\alpha} \mathbf{F}_{\nu\mu}^{(d)\alpha\beta} \tilde{\Psi}_{\alpha\beta} d\mathbf{R}_{\alpha\beta} d\mathbf{Q}^\alpha d\mathbf{Q}^\beta$$

(8.21) $$\mathbf{q}^{(d)} = \tfrac{1}{2} \sum_{\alpha\beta\nu\mu} \iiint \mathbf{R}_{\mu\nu}^{\beta\alpha} \mathbf{F}_{\nu\mu}^{(d)\alpha\beta} \cdot [[\dot{\mathbf{r}}_\nu^\alpha - \mathbf{v}]]^{\alpha\beta} \tilde{\Psi}_{\alpha\beta} d\mathbf{R}_{\alpha\beta} d\mathbf{Q}^\alpha d\mathbf{Q}^\beta$$

Table 2. General flux contributions for polymeric liquids*

Kinetic contributions:

(6.10) $$\mathbf{j}_\alpha = \sum_{\nu} m_\nu^\alpha \int [[\dot{\mathbf{r}}_\nu^\alpha - \mathbf{v}]]^\alpha \Psi_\alpha(\mathbf{r} - \mathbf{R}_\nu^\alpha, \mathbf{Q}^\alpha, t)\, d\mathbf{Q}^\alpha$$

(7.8) $$\boldsymbol{\pi}^{(k)} = \sum_{\alpha\nu} m_\nu^\alpha \int [[(\dot{\mathbf{r}}_\nu^\alpha - \mathbf{v})(\dot{\mathbf{r}}_\nu^\alpha - \mathbf{v})]]^\alpha \Psi_\alpha(\mathbf{r} - \mathbf{R}_\nu^\alpha, \mathbf{Q}^\alpha, t)\, d\mathbf{Q}^\alpha$$

(8.11) $$\begin{aligned} \mathbf{q}^{(k)} = &\sum_{\alpha\nu} \tfrac{1}{2} m_\nu^\alpha \int [[(\dot{\mathbf{r}}_\nu^\alpha - \mathbf{v}) \cdot (\dot{\mathbf{r}}_\nu^\alpha - \mathbf{v})(\dot{\mathbf{r}}_\nu^\alpha - \mathbf{v})]]^\alpha \Psi_\alpha(\mathbf{r} \quad \mathbf{R}_\nu^\alpha, \mathbf{Q}^\alpha, t)\, d\mathbf{Q}^\alpha \\ &+ \tfrac{1}{2} \sum_{\alpha\nu\mu} \int \phi_{\nu\mu}^\alpha [[\dot{\mathbf{r}}_\nu^\alpha - \mathbf{v}]]^\alpha \Psi_\alpha(\mathbf{r} - \mathbf{R}_\nu^\alpha, \mathbf{Q}^\alpha, t)\, d\mathbf{Q}^\alpha \\ &+ \tfrac{1}{2} \sum_{\alpha\beta\nu\mu} \iiint \phi_{\nu\mu}^{(d)\alpha\beta} [[\dot{\mathbf{r}}_\nu^\alpha - \mathbf{v}]]^{\alpha\beta} \\ &\qquad \cdot \tilde{\Psi}_{\alpha\beta}\left(\mathbf{r} - \mathbf{R}_\nu^\alpha + \frac{m_m^\beta}{m_m^\alpha + m_m^\beta} \mathbf{R}_{\alpha\beta}, \mathbf{R}_{\alpha\beta}, \mathbf{Q}^\alpha, \mathbf{Q}^\beta, t\right) d\mathbf{R}_{\alpha\beta}\, d\mathbf{Q}^\alpha\, d\mathbf{Q}^\beta \end{aligned}$$

External force contributions:

(7.12) $$\boldsymbol{\pi}^{(e)} = \sum_{\alpha\nu} \iint_0^1 \mathbf{R}_\nu^\alpha \mathbf{F}_\nu^{(e)\alpha} \Psi_\alpha(\mathbf{r} - \xi \mathbf{R}_\nu^\alpha, \mathbf{Q}^\alpha, t)\, d\xi\, d\mathbf{Q}^\alpha$$

(8.14) $$\mathbf{q}^{(e)} = \sum_{\alpha\nu} \iint_0^1 \mathbf{R}_\nu^\alpha \mathbf{F}_\nu^{(e)\alpha} \cdot [[\dot{\mathbf{r}}_\nu^\alpha - \mathbf{v}]]^\alpha \Psi_\alpha(\mathbf{r} - \xi \mathbf{R}_\nu^\alpha, \mathbf{Q}^\alpha, t)\, d\xi\, d\mathbf{Q}^\alpha$$

Intramolecular force contributions:

(7.15) $$\boldsymbol{\pi}^{(\phi)} = \sum_{\alpha\nu} \iint_0^1 \mathbf{R}_\nu^\alpha \mathbf{F}_\nu^{(\phi)\alpha} \Psi_\alpha(\mathbf{r} - \xi \mathbf{R}_\nu^\alpha, \mathbf{Q}^\alpha, t)\, d\xi\, d\mathbf{Q}^\alpha$$

(8.18) $$\mathbf{q}^{(\phi)} = \tfrac{1}{2} \sum_{\alpha\nu\mu} \iint_0^1 \mathbf{R}_\nu^\alpha \mathbf{F}_{\nu\mu}^{(\phi)\alpha} \cdot [[(\dot{\mathbf{r}}_\nu^\alpha - \mathbf{v}) + (\dot{\mathbf{r}}_\mu^\alpha - \mathbf{v})]]^\alpha\, \Psi_\alpha(\mathbf{r} - \xi \mathbf{R}_\nu^\alpha, \mathbf{Q}^\alpha, t)\, d\xi\, d\mathbf{Q}^\alpha$$

* In all these expressions it is understood that $\mathbf{v}$ is a function of $\mathbf{r}$ and t.

amining polymer-solvent interactions. It is hoped that the results in Tables 1 and 2 will provide the basis for fruitful exploration of these problems as well as completely new applications.

2 Notational Preliminaries

In describing flowing polymeric liquids it is probably not feasible to use detailed models that describe the locations of all the atoms in the polymer molecules. Consequently, it is necessary to use some kind of mechanical models that portray the overall molecular architecture. Bead-spring models have been widely used with considerable success for relating macroscopic properties to the main features of the molecular architecture. Even the simplest of these models – the elastic dumbbell models – are capable of describing polymer orientation and polymer stretching. More complicated chain, ring, and star models reflect better the molecular structure and allow for the portrayal of the most important internal molecular motions as well.

In this section we introduce the notation used for describing the bead-spring models. First we set forth the notation needed for describing the location and momenta for the beads. Then we discuss the symbols needed for describing the various potential energies and forces in the system.

2.1 Coordinates and Momenta for Bead-Spring Models (DPL, Sects. 11.6 and 15.1)

We consider polymer models made up of N "beads," joined together in some way by "springs"; some examples for $N = 5$ are shown in Fig. 1. Various kinds of springs have been used, including the following:

(A) Hookean (or Gaussian) $\phi = \frac{1}{2}HQ^2$

(B) Frenkel $\phi = \frac{1}{2}H(Q - Q_0)^2$

[Q_0 = length in the absence of tension in spring]

(C) Warner (or "FENE") $\phi = -\frac{1}{2}HQ_0^2 \ln\left[1 - \left(\frac{Q}{Q_0}\right)^2\right]$

[Q_0 = maximum extended length of spring]

Here ϕ is the potential energy of the spring, and Q is the interbead distance. Hookean springs have been used largely because many analytical expressions can be obtained; they are used, e.g., in the Rouse and Zimm chain models. The Frenkel springs, with a large spring constant, are useful for describing models with almost rigid bonds or almost rigid angles. The finitely extensible, nonlinear elastic (Warner) springs are useful since they approximate roughly the inverse Langevin springs derived from statistical considerations. The physical origin of the bead-spring models is discussed in Chap. 11 of DPL.

One might think that one could get statistical mechanical results for a bead-rod chain from those for a bead-spring chain with Frenkel connectors by letting the spring constant H in the latter go to infinity. It was shown by Hassager [10a] that the results for a bead-rod chain model differ from those based on a "stiffened spring model", and this difference was further substantiated by Gottlieb [10b] using molecular dynamics methods. The lack of uniqueness in the spring-stiffening procedure was further discussed by van Kampen [10c].

If desired, "internal viscosity" can be included by incorporating dashpots in parallel with the springs in the mechanical models. However, this introduces nonconservative forces into the problem. To avoid dealing with nonconservative forces, one can use momentum-space averages of the dashpot term as explained in DPL, Eq. 13C.2–2. We know of no molecular theory that can legitimize the use of dashpots in molecular models.

The beads in the mechanical models are numbered in an arbitrary manner. We let the position vector of bead ν with respect to an arbitrary laboratory-fixed

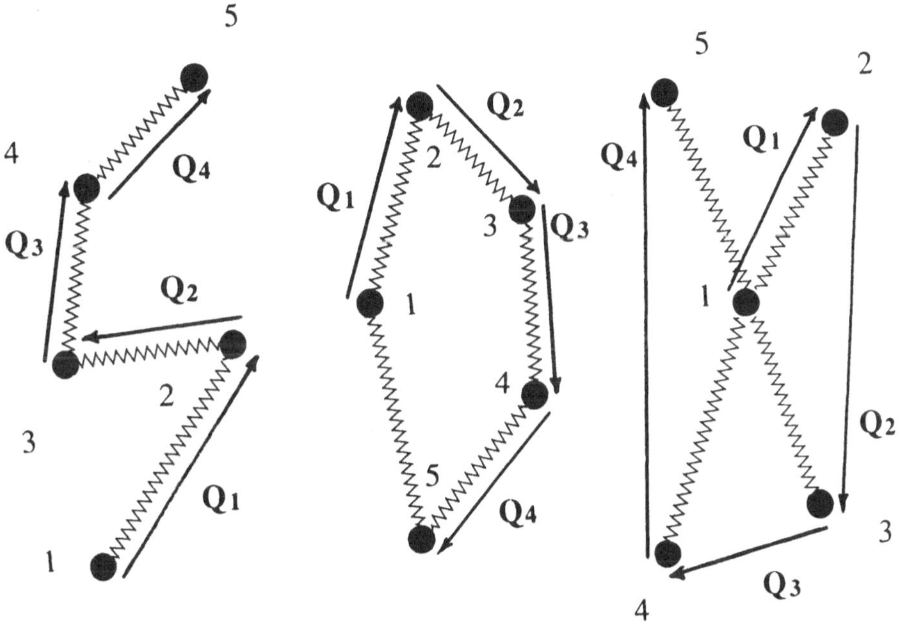

Fig. 1. Bead-spring models with $N = 5$, showing how the vectors $\mathbf{Q}_k$ are chosen

coordinate system be $\mathbf{r}_\nu$ ($\nu = 1, 2, \ldots N$), and its mass be m_ν. We then define the center-of-mass position vector $\mathbf{r}_c$ and the relative position vectors $\mathbf{Q}_k$ ($k = 1, 2, \ldots N - 1$) by:

$$\mathbf{r}_c = \Sigma_\nu m_\nu \mathbf{r}_\nu / \Sigma_\nu m_\nu = (1/m_m) \Sigma_\nu m_\nu \mathbf{r}_\nu \tag{2.1}$$

$$\mathbf{Q}_k = \mathbf{r}_{k+1} - \mathbf{r}_k = \sum_{\nu=1}^{N} \bar{B}_{k\nu} \mathbf{r}_\nu \tag{2.2}$$

in which m_m is the mass of the molecule. The $\mathbf{r}_\nu$ may be expressed in terms of $\mathbf{r}_c$ and the $\mathbf{Q}_k$ thus:

$$\mathbf{r}_\nu = \mathbf{r}_c + \sum_{k=1}^{N-1} B_{\nu k} \mathbf{Q}_k \tag{2.3}$$

It is sometimes convenient to introduce position vectors referred to the center of mass:

$$\mathbf{R}_\nu = \mathbf{r}_\nu - \mathbf{r}_c = \sum_{k=1}^{N-1} B_{\nu k} \mathbf{Q}_k \tag{2.4}$$

It is important to note that $\mathbf{R}_\nu$ is a function of the $\mathbf{Q}_k$ and furthermore that $\Sigma_\nu m_\nu \mathbf{R}_\nu = \mathbf{0}$. As is shown in Figure 1, the $\mathbf{Q}_k$ vectors are chosen in such a way that they form a "chain"; this means that the $\mathbf{Q}_k$ vectors are not necessarily coincident with the springs.

We use dots to indicate time derivatives: $\dot{\mathbf{r}}_\nu$, $\dot{\mathbf{R}}_\nu$, $\dot{\mathbf{r}}_c$, and $\dot{\mathbf{Q}}_k$. The momentum of a bead is given by

$$\mathbf{p}_\nu = m_\nu \dot{\mathbf{r}}_\nu \tag{2.5}$$

In Eqs. (2.2) and (2.3) we have introduced the non-square matrices

$$\bar{B}_{k\nu} = \delta_{k+1,\nu} - \delta_{k\nu} \tag{2.6}$$

$$\begin{cases} B_{\nu k} = (1/m_m) \sum_{\mu=1}^{k} m_\mu & \text{for } k < \nu \\ B_{\nu k} = -(1/m_m) \sum_{\mu=k+1}^{N} m_\mu & \text{for } k \geq \nu \end{cases} \tag{2.7}$$

in which $\delta_{k\nu}$ is the Kronecker delta. The $B_{\nu k}$ here are generalizations of the $B_{\nu k}$ defined in Eq. (11.6–6) of DPL, where all bead masses of one molecule were taken to be the same. The elements of these matrices are constants, and they satisfy the following relations:

$$\Sigma_\nu \bar{B}_{k\nu} = 0 \tag{2.8}$$

$$\Sigma_\nu m_\nu B_{\nu k} = 0 \tag{2.9}$$

$$\Sigma_\nu \bar{B}_{j\nu} B_{\nu k} = \delta_{jk} \tag{2.10}$$

$$\Sigma_k B_{\nu k} \bar{B}_{k\mu} = \delta_{\nu\mu} - (m_\mu/m_m) \tag{2.11}$$

Two other matrices are often used: the *Rouse matrix* with elements A_{jk} and the *Kramers matrix* with elements C_{jk}. These matrices are defined as follows:

$$A_{jk} = \Sigma_\nu \bar{B}_{j\nu} \bar{B}_{k\nu} \quad \text{and} \quad C_{jk} = \Sigma_\nu B_{\nu j} B_{\nu k} \tag{2.12}$$

and their eigenvalues are designated by a_j and c_j respectively. For the simplest elastic dumbbell model (two beads, both with mass m, and one connecting spring) these quantities are:

$$\begin{array}{lll} \bar{B}_{11} = -1 & B_{11} = -\tfrac{1}{2} & A_{11} = a_1 = 2 \\ \bar{B}_{12} = +1 & B_{21} = +\tfrac{1}{2} & C_{11} = c_1 = \tfrac{1}{2} \end{array}$$

These quantities are encountered in Sects. 13–16.

Finally we give the following chain-rule relation among the derivatives, which is used later:

$$\frac{\partial g}{\partial \mathbf{r}_\nu} = \frac{m_\nu}{m_m}\frac{\partial f}{\partial \mathbf{r}_c} + \sum_k \bar{B}_{k\nu} \frac{\partial f}{\partial \mathbf{Q}_k} \tag{2.13}$$

in which $f(\mathbf{r}_c, \mathbf{Q}_1, \mathbf{Q}_2, \ldots \mathbf{Q}_{N-1}, t) = g(\mathbf{r}_1, \mathbf{r}_2, \ldots \mathbf{r}_N, \mathbf{t})$. We also note the following Jacobian of transformation:

$$\left| \frac{\partial(\mathbf{r}_1, \mathbf{r}_2, \ldots \mathbf{r}_N)}{\partial(\mathbf{r}_c, \mathbf{Q}_1, \ldots \mathbf{Q}_{N-1})} \right| = 1 \tag{2.14}$$

This is needed in changing variables in configuration space integrals.

Up to this point we have considered one polymer molecule only. In a mixture of several different chemical species it is necessary to attach a superscript α to denote the species; thus, corresponding to $\mathbf{r}_\nu$, $\mathbf{r}_c$, $\mathbf{Q}_k$, $\mathbf{R}_\nu$ and m_m, m_ν, $\bar{B}_{k\nu}$, $B_{\nu k}$, we will write $\mathbf{r}_\nu^\alpha$, $\mathbf{r}_c^\alpha$, $\mathbf{Q}_k^\alpha$, $\mathbf{R}_\nu^\alpha$ and m_m^α, m_ν^α, $\bar{B}_{k\nu}^\alpha$, $B_{\nu k}^\alpha$ for quantities associated with a molecule of species α. We will also use the notation N_α for the number of beads in molecules of species α. The various chemical species, α, may be solvent molecules or polymer molecules. If one species is the solvent, we may label it "s".

In some instances one may wish to study polydisperse systems. These may be treated as mixtures, in which the various "species" have the same general chemical structure but differ only in the molecular weight (that is, the number of beads N_α).

The polymeric liquid, however, consists of many molecules of each species α. For quantities associated with the ith molecule of species α, we use superscripts αi; thus we write $\mathbf{r}_\nu^{\alpha i}$, $\mathbf{r}_c^{\alpha i}$, $\mathbf{Q}_k^{\alpha i}$, $\mathbf{R}_\nu^{\alpha i}$. The quantities m_m^α, m_ν^α, $\bar{B}_{k\nu}^\alpha$, $B_{\nu k}^\alpha$ do not need an index i, since they are the same for all molecules i of species α. Figure 2 shows a portion of a three-component system.

Sometimes we need to have symbols for a collection of coordinates or momenta. These sets of coordinates are indicated as follows:

$\mathbf{r}^\alpha$ = the set of all N_α coordinates $\mathbf{r}_\nu^\alpha$ for the beads of a molecule of species α (2.15)

$\mathbf{p}^\alpha$ = the set of all N_α momenta $\mathbf{p}_\nu^\alpha$ for the beads of a molecule of species α (2.16)

$\mathbf{Q}^\alpha$ = the set of all $N_\alpha - 1$ relative position vectors $\mathbf{Q}_k^\alpha$ for a molecule of species α (2.17)

and similarly

$\mathbf{r}^{\alpha i}$ = the set of all N_α coordinates $\mathbf{r}_\nu^{\alpha i}$ for the beads of the ith molecule of species α (2.18)

$\mathbf{p}^{\alpha i}$ = the set of all N_α momenta $\mathbf{p}_\nu^{\alpha i}$ for the beads of the ith molecule of species α (2.19)

We also use the notation $d\mathbf{r}^\alpha$ as a shorthand for $d\mathbf{r}_1^\alpha d\mathbf{r}_2^\alpha \cdots d\mathbf{r}_{N_\alpha}^\alpha$, and $d\mathbf{Q}^\alpha$ as an abbreviation for $d\mathbf{Q}_1^\alpha d\mathbf{Q}_2^\alpha \cdots d\mathbf{Q}_{N_\alpha - 1}^\alpha$.

A set of symbols is also needed for vectors between beads within one molecule or between beads of different molecules:

$\mathbf{R}_{\nu\mu}^{\alpha i} = \mathbf{r}_\mu^{\alpha i} - \mathbf{r}_\nu^{\alpha i}$ = vector from bead ν to bead μ within the ith molecule of species α (2.20)

$\mathbf{R}_{\nu\mu}^{\alpha i,\beta j} = \mathbf{r}_\mu^{\beta j} - \mathbf{r}_\nu^{\alpha i}$ = vector from bead ν of the ith molecule of species α to bead μ of the jth molecule of species β (2.21)

with similar notation without the i's and j's, to designate interbead vectors between beads of a molecule of species α, or between the beads of two molecules α and β.

Following the same pattern, we use the following notation for the vector between the centers of mass of two molecules:

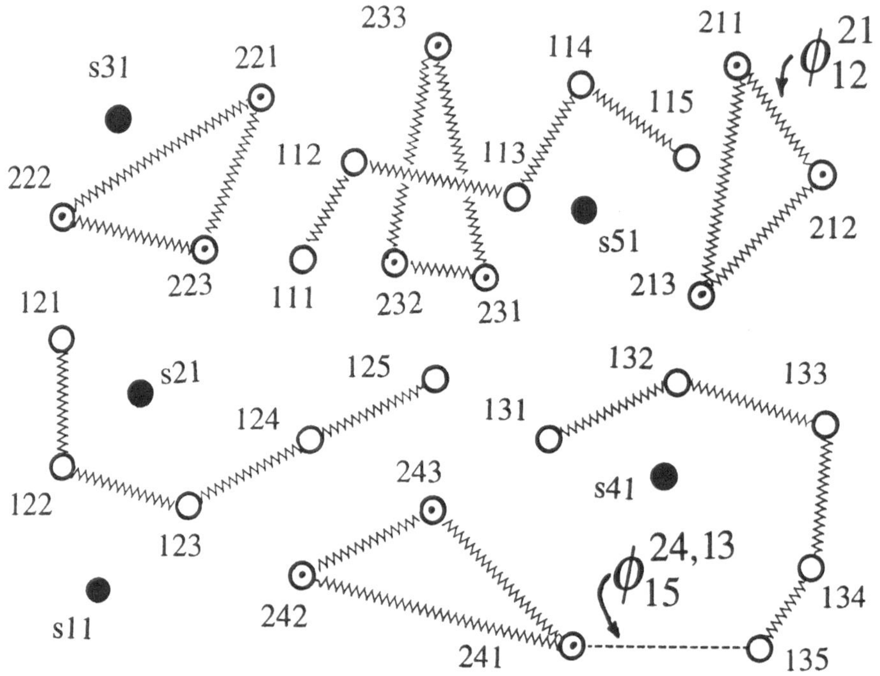

Fig. 2. A polymer solution with solvent $\alpha = s$ and two polymer solutes $\alpha = 1$, $\alpha = 2$. The number triples adjacent to each bead are $\alpha i v$(α chemical species, i = number of the molecule, v = number of bead).

α	s	1	2
N_α	1	5	3

ϕ_{12}^{21} is the potential energy for beads "1" and "2" of molecule "21," and $\phi_{15}^{24,13}$ is the potential energy for bead "1" of molecule "24" and bead "5" of molecule "13"

$$\mathbf{R}_{\alpha\beta} = \mathbf{r}_c^\beta - \mathbf{r}_c^\alpha = \text{vector from the center of mass of a molecule of species } \alpha \text{ to the center of mass of a molecule of species } \beta \tag{2.22}$$

Reversing the order of the indices for the interbead and intermolecule vectors changes the sign; for example, $\mathbf{R}_{\alpha\beta} = -\mathbf{R}_{\beta\alpha}$.

Finally we define:

$$\mathbf{r}_c^{\alpha\beta} = \frac{m_m^\alpha \mathbf{r}_c^\alpha + m_m^\beta \mathbf{r}_c^\beta}{m_m^\alpha + m_m^\beta} = \text{center of mass of a pair of molecules } \alpha \text{ and } \beta \tag{2.23}$$

This completes the definitions of position vectors, momenta, and other quantities needed in the development.

It is assumed that the beads of different molecules interact with one another according to a Lennard-Jones type of force, attractive at large distances and repulsive at small distances. Thus, the interbead force vector is taken to be collinear with the interbead vector.

As suggested by Fig. 1, the interbead forces within a single molecule are described by springs. However, if it is desired to take into account the "excluded volume" effect (the fact that the various segments of the macromolecule have finite volume and hence cannot overlap), then one can in addition include a Lennard-Jones type of interaction between those beads that are not connected by springs. Here again, the interbead force vector is taken to be collinear with the interbead vector.

2.2 Potentials and Forces for Bead-Spring Models (DPL, Sect. 16.1c)

In the foregoing subsection we gave the notation for the bead locations and masses. Here we turn to the potential energies describing the forces between beads, including the "spring forces" (and excluded volume forces) within a molecule and the "Lennard-Jones type forces" between molecules:

Symbol:	*Potential energy of:*	*A function of:*
$\phi^{\alpha i}$	A single molecule, αi (intramolecular contribution only)	$\mathbf{r}_\nu^{\alpha i}$ $(\nu = 1, 2, ..N_\alpha)$
$\phi^{(e)\alpha i}$	A single molecule, αi, (external field contribution only)	$\mathbf{r}_\nu^{\alpha i}$ $(\nu = 1, 2, \ldots N_\alpha)$
Φ	All the molecules in the liquid (intramolecular and intermolecular, but *not* external field, contributions)	$\mathbf{r}_\nu^{\alpha i}$ $(\nu = 1, 2, \ldots N_\alpha;$ $\alpha = 1, 2, 3, \ldots;$ $i = 1, 2, 3 \ldots)$

The above potential energies are assumed to be sums of pairwise interactions as follows:[1]

$$\phi^{\alpha i} = \tfrac{1}{2}\sum_\nu \sum_\mu \phi_{\nu\mu}^{\alpha i} \quad \text{in which, if } \nu \neq \mu,\ \phi_{\nu\mu}^{\alpha i} \text{ depends only on}$$

$$R_{\nu\mu}^{\alpha i} = |\mathbf{R}_{\nu\mu}^{\alpha i}| = |\mathbf{r}_\mu^{\alpha i} - \mathbf{r}_\nu^{\alpha i}|;\ \text{if } \nu = \mu, \text{ then } \phi_{\nu\mu}^{\alpha i} = 0 \quad (2.24)$$

$$\phi^{(e)\alpha i} = \sum_\nu \phi_\nu^{(e)\alpha i} \text{ in which } \phi_\nu^{(e)\alpha i} \text{ depends on only one } \mathbf{r}_\nu^{\alpha i} \quad (2.25)$$

[1] The factors of $\frac{1}{2}$ appearing in Eqs (2.24), (2.26), and (2.27) are included to avoid counting interactions twice. The potential energies for pairs of beads are symmetric in the indices: $\phi_{\nu\mu}^{\alpha i} = \phi_{\mu\nu}^{\alpha i}$, and $\phi_{\nu\mu}^{\alpha i,\beta j} = \phi_{\mu\nu}^{\beta j,\alpha i}$. Note further that:

$$\frac{\partial}{\partial \mathbf{r}_\nu^{\alpha i}}\Phi = \frac{\partial}{\partial \mathbf{r}_\nu^{\alpha i}} \sum_{\beta j\mu} \sum_{\gamma k\eta} \frac{1}{2} \phi_{\mu m}^{\beta j,\gamma k}(|\mathbf{r}_\mu^{\beta j} - \mathbf{r}_\eta^{\gamma k}|)$$

$$= \frac{\partial}{\partial \mathbf{r}_\nu^{\alpha i}} \frac{1}{2}\left(\sum_{\gamma k\eta} \phi_{\nu\eta}^{\alpha i,\gamma k} + \sum_{\beta j\mu} \phi_{\mu\nu}^{\beta j,\alpha i}\right) = \frac{\partial}{\partial \mathbf{r}_\nu^{\alpha i}} \sum_{\beta j\mu} \phi_{\nu\mu}^{\alpha i,\beta j} \quad (2.26a)$$

This should make it clear why there is no factor of $\frac{1}{2}$ appearing in Eqs. (2.28), (2.30), and (2.31).

$$\Phi = \tfrac{1}{2}\sum_{\alpha i\nu}\sum_{\beta j\mu} \phi_{\nu\mu}^{\alpha i,\beta j}$$ in which the $\phi_{\nu\mu}^{\alpha i,\beta j}$ depends only on $R_{\nu\mu}^{\alpha i,\beta j} = |\mathbf{R}_{\nu\mu}^{\alpha i,\beta j}| = |\mathbf{r}_{\mu}^{\beta j} - \mathbf{r}_{\nu}^{\alpha i}|$; the $\phi_{\nu\nu}^{\alpha i,\alpha i}$ are defined to be zero and $\phi_{\nu\mu}^{\alpha i,\alpha i} = \phi_{\nu\mu}^{\alpha i}$ (2.26)

Sometimes it is convenient to write Φ as the sum of the intramolecular and the intermolecular contributions:

$$\Phi = \tfrac{1}{2}\sum_{\nu}\sum_{\mu} \phi_{\nu\mu}^{\alpha i} + \tfrac{1}{2}\sum_{\alpha i\nu}\sum_{\beta j\mu} \phi_{\nu\mu}^{(d)\alpha i,\beta j} \tag{2.27}$$

That is, the $\phi_{\nu\mu}^{\alpha i}$ terms describe the bead-bead interactions within a *single* molecule, whereas the $\phi_{\nu\mu}^{(d)\alpha i,\beta j}$ terms describe the bead-bead interactions between *different* molecules. Here $\phi_{\nu\mu}^{(d)\alpha i,\beta j}$ differs from $\phi_{\nu\mu}^{\alpha i,\beta j}$, because for the former, if $\alpha = \beta$, those terms with $i = j$ are zero.

We now define the following forces on bead ν of molecule αi, corresponding to the potential energies defined in Eqs (2.1), (2.2), and (2.3):

$$\mathbf{F}_{\nu}^{(\phi)\alpha i} = -\frac{\partial}{\partial \mathbf{r}_{\nu}^{\alpha i}}\phi^{\alpha i} = -\frac{\partial}{\partial \mathbf{r}_{\nu}^{\alpha i}}\sum_{\mu}\phi_{\nu\mu}^{\alpha i} = \sum_{\mu}\mathbf{F}_{\nu\mu}^{(\phi)\alpha i}$$

$$= \text{force resulting from "springs" within one molecule} \tag{2.28}$$

$$\mathbf{F}_{\nu}^{(e)\alpha i} = -\frac{\partial}{\partial \mathbf{r}_{\nu}^{\alpha i}}\phi^{(e)\alpha i} = -\frac{\partial}{\partial \mathbf{r}_{\nu}^{\alpha i}}\phi_{\nu}^{(e)\alpha i}$$

$$= \text{force exerted by an external field} \tag{2.29}$$

$$\mathbf{F}_{\nu}^{(\Phi)\alpha i} = -\frac{\partial}{\partial \mathbf{r}_{\nu}^{\alpha i}}\Phi = -\frac{\partial}{\partial \mathbf{r}_{\nu}^{\alpha i}}\sum_{\beta j\mu}\phi_{\nu\mu}^{\alpha i,\beta j} = \sum_{\beta j\mu}\mathbf{F}_{\nu\mu}^{(\Phi)\alpha i,\beta j}$$

$$= \text{force due to all other "beads" in the liquid } (\alpha i\nu \neq \beta j\mu) \tag{2.30}$$

In addition we shall need an expression for the force on bead ν of one molecule due to the interactions with all beads on all other molecules

$$\mathbf{F}_{\nu}^{(d)\alpha i} = -\frac{\partial}{\partial \mathbf{r}_{\nu}^{\alpha i}}\sum_{\beta j\mu}\phi_{\nu\mu}^{(d)\alpha i,\beta j} = \sum_{\beta j\mu}\mathbf{F}_{\nu\mu}^{(d)\alpha i,\beta j} \tag{2.31}$$

The total force on bead ν of molecule αi can then be written as:

$$\mathbf{F}_{\nu}^{\alpha i} = \mathbf{F}_{\nu}^{(\Phi)\alpha i} + \mathbf{F}_{\nu}^{(e)\alpha i} \quad \text{or} \quad \mathbf{F}_{\nu}^{\alpha i} = \mathbf{F}_{\nu}^{(\phi)\alpha i} + \mathbf{F}_{\nu}^{(d)\alpha i} + \mathbf{F}_{\nu}^{(e)\alpha i} \tag{2.32}$$

and, according to Newton's second law of motion:

$$\mathbf{F}_{\nu}^{\alpha i} = \dot{\mathbf{p}}_{\nu}^{\alpha i} \quad \text{and} \quad \mathbf{p}_{\nu}^{\alpha i} = m_{\nu}^{\alpha}\dot{\mathbf{r}}_{\nu}^{\alpha i} \tag{2.33}$$

Because of the second of these relations we shall switch freely from momenta to velocities in subsequent derivations.

Similarly, we define symbols for forces on the center of mass of αi, for example:

$$\mathbf{F}^{(e)\alpha i} = -\frac{\partial}{\partial \mathbf{r}_{c}^{\alpha i}}\phi^{(e)\alpha i} = \text{force due to external field} \tag{2.34}$$

$$\mathbf{F}^{(d)\alpha i} = -\frac{\partial}{\partial \mathbf{r}_c^{\alpha i}} \Phi = \text{force due to all other molecules} \tag{2.35}$$

In these expressions it is implied that there has been a change of variables from $\mathbf{r}_\nu^{\alpha i}$ to $\mathbf{r}_c^{\alpha i}$, $\mathbf{Q}_k^{\alpha i}$.

Later, in Sect. 11, notation is introduced for Brownian forces and hydrodynamic forces.

3 The Liouville Equation and the General Equation of Change (DPL, Sect. 17.2)

The instantaneous dynamical state of the polymeric liquid mixture (the "system" under consideration) is described by a "system point" in the complete phase space, that is, by the values of all the $\mathbf{r}_\nu^{\alpha i}$ and $\mathbf{p}_\nu^{\alpha i}$ for $\alpha = 0, 1, 2, \ldots$; $i = 1, 2, \ldots$; $\nu = 1, 2, \ldots N_\alpha$. For brevity we will sometimes use the symbols r, p, and x to designate, respectively, the complete set of position coordinates, the complete set of momentum coordinates, and the complete set of phase space coordinates.

We consider an ensemble of identical isolated systems. The state of the ensemble is described by a distribution function $f(x, t)$ in the system phase space; $f(x, t)$ is taken to be normalized to unity:

$$\int f(x, t)dx = 1 \tag{3.1}$$

The average value of a function $B(x)$ is then given by

$$\int B(x) f(x, t)dx = \langle B \rangle \tag{3.2}$$

The distribution function in the configuration space $\Psi(r, t)$ is defined as follows:

$$\int f(x, t)dp = \Psi(r, t) \tag{3.3}$$

and the average value of a function $B(r)$ in the configuration space is given by

$$\int B(r) f(x, t)dx = \int B(r)\Psi(r, t)dr = \langle B \rangle \tag{3.4}$$

Note that the brackets $\langle \cdots \rangle$ in both Eq. (3.2) and Eq. (3.4) are averages over the entire system phase space.

The distribution function f satisfies an equation of continuity in the phase space (Ref. 12, p. 40):

$$\frac{\partial}{\partial t} f = -\sum_{\alpha i \nu} \left(\frac{\partial}{\partial \mathbf{r}_\nu^{\alpha i}} \cdot \dot{\mathbf{r}}_\nu^{\alpha i} f + \frac{\partial}{\partial \mathbf{p}_\nu^{\alpha i}} \cdot \dot{\mathbf{p}}_\nu^{\alpha i} f \right) \tag{3.5}$$

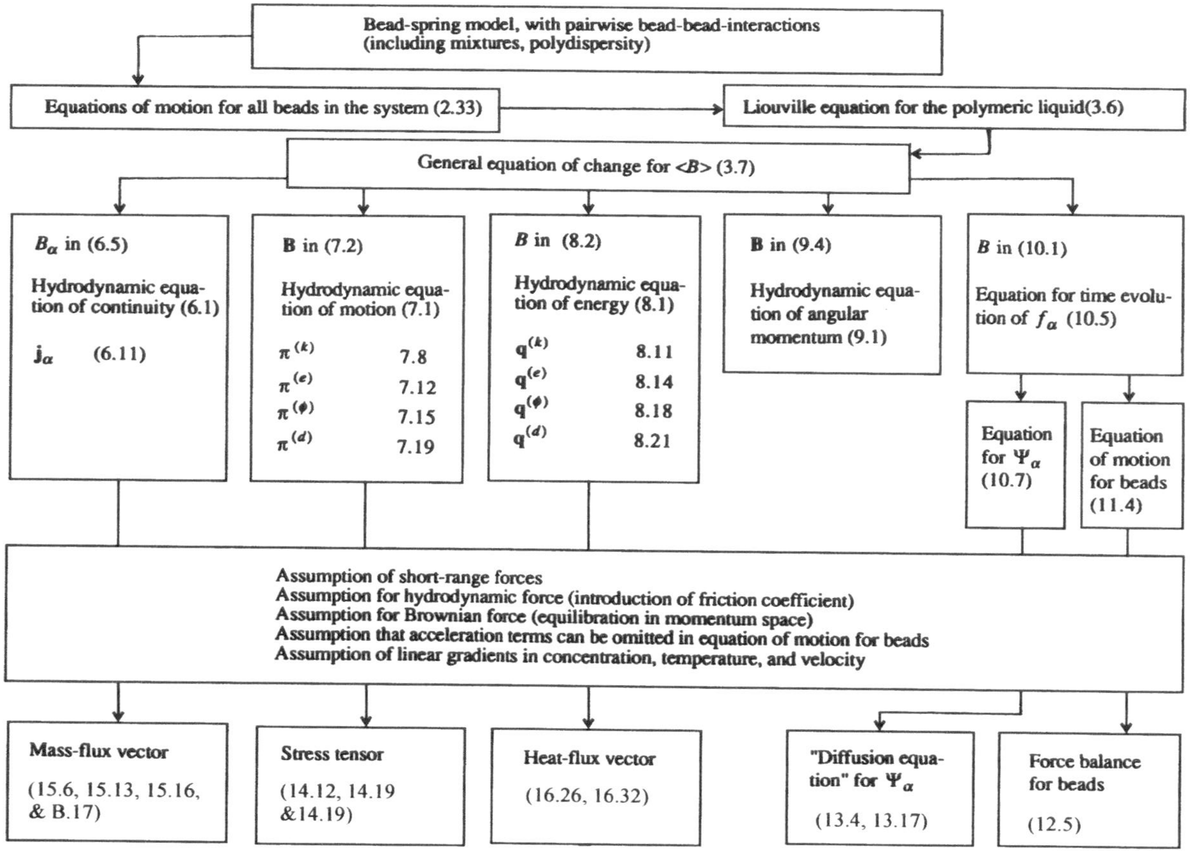

Fig. 3. Diagram showing how the equations of change and the flux expressions are obtained from the "general equation of change," and how the five main assumptions are used to get expressions for the fluxes for special molecular models

Then use of Newton's second law of motion, Eq (2.33), leads to the *Liouville equation*[2]

$$\frac{\partial}{\partial t} f = -\sum_{\alpha i \nu} \left(\frac{\mathbf{p}_\nu^{\alpha i}}{m_\nu^\alpha} \cdot \frac{\partial}{\partial \mathbf{r}_\nu^{\alpha i}} f + \mathbf{F}_\nu^{\alpha i} \cdot \frac{\partial}{\partial \mathbf{p}_\nu^{\alpha i}} f \right) \equiv -\mathbb{L} f \tag{3.6}$$

in which $\mathbb{L}$ is the *Liouville operator*. When Eq. (3.6) is multiplied by B and integrated over the system phase space (see, for example, [6]), we get

$$\frac{\partial}{\partial t} \langle B \rangle = \langle \mathbb{L} B \rangle \tag{3.7}$$

which is the *general equation of change*. This equation plays a central role in the kinetic theory. By making various choices of B we can get:

Sect. 6: the hydrodynamic equation of continuity for each species and a formal expression for the mass flux vector of each species
Sect. 7: the hydrodynamic equation of motion for the liquid mixture and a formal expression for the stress tensor
Sect. 8: the energy equation for the liquid and a formal expression for the heat flux vector
Sect. 9: the angular momentum equation
Sect. 10: an equation for the time evolution of the phase space distribution function for each species; from this equation one can get the equations of motion for the beads and for the molecule mass centers (see Sect. 11)

In short, one can obtain systematically from Eq (3.7) all the kinetic theory equations needed for a given molecular model (see Fig. 3).

4 Contracted Distribution Functions (DPL, Sects. 12.5 and 17.3)

In the foregoing section the distribution function in the complete phase space, $f(x, t)$, was discussed and the general equation of change presented. In the sections to follow we often present results in terms of averages with respect to lower-order distribution functions, specifically averages involving the phase space or the configuration space of one molecule or a pair of molecules. This section is devoted to the definitions of these contracted distributions functions.

---

[2] For systems with no constraints, Newton's laws of motion are the same as the Hamilton canonical equations of motion obtained from the Hamiltonian function:

$$\mathbb{H} = \sum_{\alpha i \nu} \frac{1}{2m_\nu^\alpha} (\mathbf{p}_\nu^{\alpha i} \cdot \mathbf{p}_\nu^{\alpha i}) + \sum_{\alpha i} \phi^{(e)\alpha i} + \Phi \tag{3.6a}$$

and Eq. (3.6) is the same as $\partial f/\partial t = -\{f, \mathbb{H}\}$, in which $\{f, \mathbb{H}\}$ is the *Poisson bracket* of f and $\mathbb{H}$.

4.1 Singlet Distribution Functions

The singlet phase-space distribution function $f_\alpha(\mathbf{r}^\alpha, \mathbf{p}^\alpha, t)$ is defined as:

$$f_\alpha(\mathbf{r}^\alpha, \mathbf{p}^\alpha, t) = \langle \Sigma_i \delta(\mathbf{r}_1^{\alpha i} - \mathbf{r}_1^\alpha)\delta(\mathbf{r}_2^{\alpha i} - \mathbf{r}_2^\alpha) \ldots \delta(\mathbf{p}_1^{\alpha i} - \mathbf{p}_1^\alpha)\delta(\mathbf{p}_2^{\alpha i} - \mathbf{p}_2^\alpha)\ldots\rangle$$
$$= \langle \Sigma_i \delta(\mathbf{r}^{\alpha i} - \mathbf{r}^\alpha)\delta(\mathbf{p}^{\alpha i} - \mathbf{p}^\alpha)\rangle \tag{4.1}$$

In the first line the Dirac delta function of a vector argument represents the product of the delta functions of the three Cartesian components: $\delta(\mathbf{r}) = \delta(x)\delta(y)\delta(z)$. In the second line we have introduced a shorthand for the product of N_α Dirac delta functions each with a vector argument. The interpretation of the singlet distribution function is:

$f_\alpha(\mathbf{r}_1^\alpha, \mathbf{r}_2^\alpha, \ldots \mathbf{r}_{N_\alpha}^\alpha, \mathbf{p}_1^\alpha, \mathbf{p}_2^\alpha, \ldots \mathbf{p}_{N_\alpha}^\alpha, t)d\mathbf{r}_1^\alpha d\mathbf{r}_2^\alpha \ldots d\mathbf{r}_{N_\alpha}^\alpha d\mathbf{p}_1^\alpha d\mathbf{p}_2^\alpha \ldots d\mathbf{p}_{N_\alpha}^\alpha$ = the ensemble average of the number of molecules of species α, for which bead "1" is located in the range $d\mathbf{r}_1^\alpha$ about $\mathbf{r}_1^\alpha$ with momentum in the range $d\mathbf{p}_1^\alpha$ about $\mathbf{p}_1^\alpha$, bead "2" is located in the range $d\mathbf{r}_2^\alpha$ about $\mathbf{r}_2^\alpha$ with momentum in the range $d\mathbf{p}_2^\alpha$ about $\mathbf{p}_2^\alpha$, etc.

The distribution function $f_\alpha(\mathbf{r}^\alpha, \mathbf{p}^\alpha, t)$ is normalized as follows:

$$\iint f_\alpha(\mathbf{r}^\alpha, \mathbf{p}^\alpha, t)\delta(\mathbf{r}_c^\alpha - \mathbf{r})d\mathbf{r}^\alpha d\mathbf{p}^\alpha = n_\alpha(\mathbf{r}, t) \tag{4.2}$$

where $n_\alpha(\mathbf{r}, t)$ is the ensemble average of the number density of centers-of-mass of species α at position $\mathbf{r}$ and time t. The singlet configurational distribution function $\bar{\Psi}_\alpha$ is defined by:

$$\bar{\Psi}_\alpha(\mathbf{r}^\alpha, t) = \int f_\alpha(\mathbf{r}^\alpha, \mathbf{p}^\alpha, t)d\mathbf{p}^\alpha = \langle \Sigma_i \delta(\mathbf{r}^{\alpha i} - \mathbf{r}^\alpha)\rangle \tag{4.3}$$

It is expected that $\bar{\Psi}_\alpha$ changes slowly with respect to the location of the center of mass of the molecule over a length scale comparable to that over which macroscopic quantities (concentration, velocity, and temperature) change. It is therefore convenient to introduce the variables of Eqs. (2.1) and (2.2) and to define Ψ_α as

$$\Psi_\alpha(\mathbf{r}_c^\alpha, \mathbf{Q}^\alpha, t) = \bar{\Psi}_\alpha(\mathbf{r}^\alpha, t) \tag{4.4}$$

with normalization given by:

$$\int \Psi_\alpha(\mathbf{r}_c^\alpha, \mathbf{Q}^\alpha, t)\delta(\mathbf{r}_c^\alpha - \mathbf{r})d\mathbf{r}_c^\alpha dQ^\alpha = \int \Psi_\alpha(\mathbf{r}, \mathbf{Q}^\alpha, t)d\mathbf{Q}^\alpha = n_\alpha(\mathbf{r}, t) \tag{4.5}$$

Both here and in Eq. (4.2) the delta function $\delta(\mathbf{r}_c^\alpha - \mathbf{r})$ is included in order to place the center of mass of the molecule at the location $\mathbf{r}$ in space; the center of mass position vector $\mathbf{r}_c^\alpha$ is a function of all the $\mathbf{r}_\nu^\alpha$ as given in Eq. (2.1). In the following sections the variable $\mathbf{r}$ may frequently – but not always – be interpreted as the position of the center of mass of a molecule. The functions $\bar{\Psi}_\alpha$ and Ψ_α contain the same information; the presence or absence of the overbar indicates the set of variables used to describe the molecular conformation.

4.2 Doublet Distribution Functions

The doublet distribution function (or pair distribution function) $f_{\alpha\beta}$, for a pair of two molecules α and β, is defined as:

$$f_{\alpha\beta}(\mathbf{r}^\alpha, \mathbf{p}^\alpha, \mathbf{r}^\beta, \mathbf{p}^\beta, t) = \langle \Sigma_i \Sigma_j \delta(\mathbf{r}^{\alpha i} - \mathbf{r}^\alpha)\delta(\mathbf{p}^{\alpha i} - \mathbf{p}^\alpha)\delta(\mathbf{r}^{\beta j} - \mathbf{r}^\beta)\delta(\mathbf{p}^{\beta j} - \mathbf{p}^\beta)\rangle \tag{4.6}$$

If $\alpha = \beta$, the terms in the double sum with $i = j$ are to be omitted. The interpretation of the doublet distribution function is as follows:

$f_{\alpha\beta}(\mathbf{r}_1^\alpha,\ldots\mathbf{r}_{N_\alpha}^\alpha,\ \mathbf{p}_1^\alpha,\ldots\mathbf{p}_{N_\alpha}^\alpha,\ \mathbf{r}_1^\beta,\ldots\mathbf{r}_{N_\beta}^\beta,\ \mathbf{p}_1^\beta,\ldots\mathbf{p}_{N_\beta}^\beta,\ t)d\mathbf{r}_1^\alpha,\ldots d\mathbf{r}_{N_\alpha}^\alpha d\mathbf{p}_1^\alpha,\ldots d\mathbf{p}_{N_\alpha}^\alpha\ d\mathbf{r}_1^\beta\ldots d\mathbf{r}_{N_\beta}^\beta$ $d\mathbf{p}_1^\beta,\ldots d\mathbf{p}_{N_\beta}^\beta$ = the ensemble average of the number of molecules of species α and β, such that bead "1" of a molecule of α is located within the range $d\mathbf{r}_1^\alpha$ about $\mathbf{r}_1^\alpha$ with momentum in the range $d\mathbf{p}_1^\alpha$ about $\mathbf{p}_1^\alpha$, etc., and that bead "1" of a molecule of β is located within the range $d\mathbf{r}_1^\beta$ about $\mathbf{r}_1^\beta$ with momentum in the range $d\mathbf{p}_1^\beta$ about $\mathbf{p}_1^\beta$, etc.

The doublet distribution function in configuration space is obtained by integrating the above function over all momenta:

$$\begin{aligned}\bar{\Psi}_{\alpha\beta}(\mathbf{r}^\alpha, \mathbf{r}^\beta, t) &= \int \mathbf{f}_{\alpha\beta}(\mathbf{r}^\alpha, \mathbf{p}^\alpha, \mathbf{r}^\beta, \mathbf{p}^\beta, t)d\mathbf{p}^\alpha d\mathbf{p}^\beta \\ &= \langle \Sigma_i \Sigma_j \delta(\mathbf{r}^{\alpha i} - \mathbf{r}^\alpha)\delta(\mathbf{r}^{\beta j} - \mathbf{r}^\beta)\rangle\end{aligned} \tag{4.7}$$

We can also use the variables of Eqs. (2.1) and (2.2) to define $\Psi_{\alpha\beta}$ as follows:

$$\Psi_{\alpha\beta}(\mathbf{r}_c^\alpha, \mathbf{Q}^\alpha, \mathbf{r}_c^\beta, \mathbf{Q}^\beta, t) = \bar{\Psi}_{\alpha\beta}(\mathbf{r}^\alpha, \mathbf{r}^\beta, t) \tag{4.8}$$

If the centers of masses of the two molecules α and β are quite far apart, so that the internal motions of the two molecules are uncorrelated, then the pair distribution function can be factorized:

$$\lim_{|\mathbf{r}_c^\beta - \mathbf{r}_c^\alpha| \to \infty} \Psi_{\alpha\beta} = \Psi_\alpha(\mathbf{r}_c^\alpha, \mathbf{Q}^\alpha, t)\Psi_\beta(\mathbf{r}_c^\beta, \mathbf{Q}^\beta, t) \tag{4.9}$$

Since $\Psi_{\alpha\beta}$ varies rapidly with the vector distance $\mathbf{R}_{\alpha\beta} = \mathbf{r}_c^\beta - \mathbf{r}_c^\alpha$ between the centers of mass of the two molecules, but weakly with the location of the center of mass $\mathbf{r}_c^{\alpha\beta}$ of the two-molecule pair, Eq. (1.22), it is useful to make a further change of variables, for which the Jacobian is unity:

$$\mathbf{r}_c^\alpha = \mathbf{r}_v^{\alpha\beta} - \frac{m_m^\beta}{m_m^\alpha + m_m^\beta}\mathbf{R}_{\alpha\beta} \tag{4.10}$$

$$\mathbf{r}_c^\beta = \mathbf{r}_c^{\alpha\beta} + \frac{m_m^\alpha}{m_m^\alpha + m_m^\beta}\mathbf{R}_{\alpha\beta} \tag{4.11}$$

Then we can define $\tilde{\Psi}_{\alpha\beta}$ by

$$\tilde{\Psi}_{\alpha\beta}(\mathbf{r}_c^{\alpha\beta}, \mathbf{R}_{\alpha\beta}, \mathbf{Q}^\alpha, \mathbf{Q}^\beta, t) = \Psi_{\alpha\beta}(\mathbf{r}_c^\alpha, \mathbf{Q}^\alpha, \mathbf{r}_c^\beta, \mathbf{Q}^\beta, t) \tag{4.12}$$

Note the following symmetry relation:

$$\tilde{\Psi}_{\beta\alpha}(\mathbf{r}_c^{\alpha\beta}, -\mathbf{R}_{\alpha\beta}, \mathbf{Q}^{\beta}, \mathbf{Q}^{\alpha}, t) = \tilde{\Psi}_{\alpha\beta}(\mathbf{r}_c^{\alpha\beta}, \mathbf{R}_{\alpha\beta}, \mathbf{Q}^{\alpha}, \mathbf{Q}^{\beta}, t) \tag{4.13}$$

The functions $\bar{\Psi}_{\alpha\beta}$, $\Psi_{\alpha\beta}$ and $\tilde{\Psi}_{\beta\alpha}$ all contain the same information, namely the probability of finding a pair of molecules (of the same or different species) in a given joint configuration; different symbols are used to emphasize that different functions of the (different) independent variables are involved. We do not, however, use overbar and tilde notation for potential energies and forces when changing the independent variables.

5 Average Values in Terms of Lower-Order Distribution Functions

In this section we present some formulas and notation that will be very useful in subsequent sections, where averages with respect to the full phase space distribution are reduced to averages with respect to lower-order distribution functions. By formalizing the reduction procedure some economy of presentation can be achieved. For those not particularly interested in the mathematical details, the principal results are given in Eqs. (5.9), (5.10), (5.19) and (5.20).

5.1 Averages over Momentum Space

We shall often abbreviate integrals of $b^{\alpha}(\mathbf{r}^{\alpha}, \mathbf{p}^{\alpha})$, defined in the phase space of a single molecule, as follows:

$$\int b^{\alpha}(\mathbf{r}^{\alpha}, \mathbf{p}^{\alpha}) f_{\alpha}(\mathbf{r}^{\alpha}, \mathbf{p}^{\alpha}, t) d\mathbf{p}^{\alpha} = [[b^{\alpha}]]^{\alpha} \bar{\Psi}_{\alpha}(\mathbf{r}^{\alpha}, t) \tag{5.1}$$

in which the double-bracketed quantity is a function of the same variables as those of the configuration-space distribution function that follows. The double-bracketed quantity is an average in momentum space in the sense that it can be written as $[[b^{\alpha}]]^{\alpha} = \int b^{\alpha} f_{\alpha} d\mathbf{p}^{\alpha} / \int f_{\alpha} d\mathbf{p}^{\alpha}$ by using Eq. (4.3).

Similarly we will abbreviate integrals over quantities defined in the phase space of a pair of molecules as follows:

$$\iint b^{\alpha}(\mathbf{r}^{\alpha}, \mathbf{p}^{\alpha}) f_{\alpha\beta}(\mathbf{r}^{\alpha}, \mathbf{p}^{\alpha}, \mathbf{r}^{\beta}, \mathbf{p}^{\beta}, t) d\mathbf{p}^{\alpha} d\mathbf{p}^{\beta} = [[b^{\alpha}]]^{\alpha\beta} \bar{\Psi}_{\alpha\beta}(\mathbf{r}^{\alpha}, \mathbf{r}^{\beta}, t) \tag{5.2}$$

in which the double-bracketed quantity once again is a function of the same variables as those in the configuration-space distribution function immediately following. The double-bracketed quantity can be viewed as a momentum-space average $[[b^{\alpha}]]^{\alpha\beta} = \iint b^{\alpha} f_{\alpha\beta} d\mathbf{p}^{\alpha} d\mathbf{p}^{\beta} / \iint f_{\alpha\beta} d\mathbf{p}^{\alpha} d\mathbf{p}^{\beta}$ by using Eq. (4.7).

5.2 Averages Involving Functions in the Phase Space of One Molecule Only

In developing expressions for the fluxes, averages of the following general form arise:

$$\bar{C}_\nu^\alpha = \langle \Sigma_i b_\nu^{\alpha i} C_\nu^{\alpha i} \delta(\mathbf{r}_\nu^{\alpha i} - \mathbf{r}) \rangle = \int \Sigma_i b_\nu^{\alpha i} C_\nu^{\alpha i} f(x, t) \delta(\mathbf{r}_\nu^{\alpha i} - \mathbf{r}) dx \tag{5.3}$$

in which $b_\nu^{\alpha i}$ is a function of the $\mathbf{p}_\mu^{\alpha i}$ for all μ, and $C_\nu^{\alpha i}$ is a function of the $\mathbf{r}_\mu^{\alpha i}$ for all μ. We now show how to write averages of this type in terms of the singlet distribution function.

We start by rewriting $\delta(\mathbf{r}_\nu^{\alpha i} - \mathbf{r})$ as the integral of the product of two delta functions:

$$\bar{C}_\nu^\alpha = \iint \Sigma_i b_\nu^{\alpha i} C_\nu^{\alpha i} f(x, t) \delta(\mathbf{r}_\nu^{\alpha i} - \mathbf{r}_\nu^\alpha) \delta(\mathbf{r}_\nu^\alpha - \mathbf{r}) d\mathbf{r}_\nu^\alpha dx \tag{5.4}$$

Next we introduce into the integrand the products of integrals over the delta functions $\delta(\mathbf{r}_\mu^{\alpha i} - \mathbf{r}_\mu^\alpha)$ with $\mu \neq \nu$, and also products of integrals over the delta functions $\delta(\mathbf{p}_\mu^{\alpha i} - \mathbf{p}_\mu^\alpha)$ for all μ, each integral just giving unity:

$$\bar{C}_\nu^\alpha = \iiint \Sigma_i b_\nu^{\alpha i} C_\nu^{\alpha i} \delta(\mathbf{r}^{\alpha i} - \mathbf{r}^\alpha) \delta(\mathbf{p}^{\alpha i} - \mathbf{p}^\alpha) f(x, t) \delta(\mathbf{r}_\nu^\alpha - \mathbf{r}) d\mathbf{r}^\alpha d\mathbf{p}^\alpha dx \tag{5.5}$$

Now, because of the inclusion of the delta functions in the integrand, the quantity $b_\nu^{\alpha i}$ can now be regarded as depending on the $\mathbf{p}_\mu^\alpha$ (for all μ), and hence the index i on $b_\nu^{\alpha i}$ can be omitted. (It is to be understood that b_ν^α is the same function of the $\mathbf{p}^\alpha$ as $b_\nu^{\alpha i}$ is of the $\mathbf{p}^{\alpha i}$). Similarly the quantity $C_\nu^{\alpha i}$ can now be regarded as depending on the $\mathbf{r}_\mu^\alpha$ (for all μ), and hence the index i on $C_\nu^{\alpha i}$ can also be omitted. We next interchange the order of integration in Eq. (5.5) and perform the integration over all of the x-variables. When use is made of the definition in Eq. (4.1), we get

$$\bar{C}_\nu^\alpha = \iint b_\nu^\alpha C_\nu^\alpha f_\alpha(\mathbf{r}^\alpha, \mathbf{p}^\alpha, t) \delta(\mathbf{r}_\nu^\alpha - \mathbf{r}) d\mathbf{r}^\alpha d\mathbf{p}^\alpha \tag{5.6}$$

When the integrations over all momenta are performed and the definition in Eq. (5.1) is used, this becomes:

$$\bar{C}_\nu^\alpha = \int [[b_\nu^\alpha]]^\alpha C_\nu^\alpha \Psi_\alpha(\mathbf{r}^\alpha, t) \delta(\mathbf{r}_\nu^\alpha - \mathbf{r}) d\mathbf{r}^\alpha \tag{5.7}$$

In going from Eq. (5.3) to this result, essentially all that has been done is to integrate over all the variables except for the coordinates of a single molecule; the integration over the momenta is indicated formally by the double-bracket notation. Keep in mind that both $[[b_\nu^\alpha]]^\alpha$ and $[C_\nu^\alpha]$ are functions of all the $\mathbf{r}_\nu^\alpha$.

Rather than using the $\mathbf{r}_\nu^\alpha$ as the independent variables for species α, we may switch to the center of mass position vector $\mathbf{r}_c^\alpha$ and the set of $(N_\alpha - 1)$ relative position vectors $\mathbf{Q}_k^\alpha$. By using Eqs. (2.3), (2.4), and (2.14) we get:

$$\bar{C}_\nu^\alpha = \iint [[b_\nu^\alpha]]^\alpha C_\nu^\alpha \Psi_\alpha(\mathbf{r}_c^\alpha, \mathbf{Q}^\alpha, t)\delta(\mathbf{r}_c^\alpha + \mathbf{R}_\nu^\alpha - \mathbf{r})d\mathbf{r}_c^\alpha d\mathbf{Q}^\alpha \tag{5.8}$$

When the integration over $\mathbf{r}_c^\alpha$ is performed, $[[b_\nu^\alpha]]^\alpha$, C_ν^α, and Ψ_α all become functions of $(\mathbf{r} - \mathbf{R}_\nu^\alpha)$; keep in mind that the $\mathbf{R}_\nu^\alpha$ depend on all the $\mathbf{Q}_k^\alpha$.

$$\bar{C}_\nu^\alpha = \int [[b_\nu^\alpha]]^\alpha C_\nu^\alpha \Psi_\alpha(\mathbf{r} - \mathbf{R}_\nu^\alpha, \mathbf{Q}^\alpha, t)d\mathbf{Q}^\alpha \tag{5.9}$$

Note carefully that the arguments of both $[[b_\nu^\alpha]]^\alpha$ and C_ν^α are identical to those of the distribution function Ψ_α. Since we will generally assume that Ψ_α varies slowly with the position of the center of mass (i.e., with the first independent variable), the integrand may be expanded in a Taylor series about $\mathbf{r}$. This gives:

$$\begin{aligned}\bar{C}_\nu^\alpha &= \langle \Sigma_i b_\nu^{\alpha i} C_\nu^{\alpha i} \delta(\mathbf{r}_\nu^{\alpha i} - \mathbf{r})\rangle \\ &= \int [[b_\nu^\alpha]]^\alpha C_\nu^\alpha \Psi_\alpha(\mathbf{r}, \mathbf{Q}^\alpha, t)dQ^\alpha - \nabla \cdot \int \mathbf{R}_\nu^\alpha [[b_\nu^\alpha]]^\alpha C_\nu^\alpha \Psi_\alpha(\mathbf{r}, \mathbf{Q}^\alpha, t)d\mathbf{Q}^\alpha \\ &\quad + \tfrac{1}{2}\nabla\nabla : \int \mathbf{R}_\nu^\alpha \mathbf{R}_\nu^\alpha [[b_\nu^\alpha]]^\alpha C_\nu^\alpha \Psi_\alpha(\mathbf{r}, \mathbf{Q}^\alpha, t)d\mathbf{Q}^\alpha + \cdots \end{aligned} \tag{5.10}$$

where now $[[b_\nu^\alpha]]^\alpha$ and C_ν^α (which may be scalars, vectors, or tensors) are functions of $\mathbf{r}$ and the $\mathbf{Q}_k^\alpha$. If $[[b_\nu^\alpha]]^\alpha$ and C_ν^α are vectors or tensors, then the sequence of the factors in the various terms of Eq. (5.10) must be carefully preserved.

Formally one can include all the terms in the expansion in Eq. (5.10) by writing:

$$\begin{aligned}\bar{C}_\nu^\alpha &= \langle \Sigma_i b_\nu^{\alpha i} C_\nu^{\alpha i} \delta(\mathbf{r}_\nu^{\alpha i} - \mathbf{r})\rangle = \int [[b_\nu^\alpha]]^\alpha C_\nu^\alpha \Psi_\alpha(\mathbf{r}, \mathbf{Q}^\alpha, t)d\mathbf{Q}^\alpha \\ &\quad - \nabla \cdot \int_0^1\!\!\int \mathbf{R}_\nu^\alpha [[b_\nu^\alpha]]^\alpha C_\nu^\alpha \Psi_\alpha(\mathbf{r} - \xi\mathbf{R}_\nu^\alpha, \mathbf{Q}^\alpha, t)\, d\xi\, d\mathbf{Q}^\alpha \end{aligned} \tag{5.11}$$

as described in Sect. 2 of DPL. Equations (5.9) to (5.11) are the main results of this subsection; they are used repeatedly in subsequent sections. The higher terms in the Taylor series in Eq. (5.11) may be needed for the description of nonhomogeneous systems and wall effects.

5.3 Averages Involving Functions in the Phase Space of Two Molecules Only

Another relevant set of average quantities involves interactions between pairs of beads belonging to different molecules. Specifically we will be concerned with averages of the form:

$$\bar{C}_\nu^{\alpha\beta} = \langle \Sigma_i \Sigma_j' b_\nu^{\alpha i} C_\nu^{\alpha i,\beta j} \delta(\mathbf{r}_\nu^{\alpha i} - \mathbf{r}) \rangle \tag{5.12}$$

in which the prime on the summation symbol indicates that $\alpha i \neq \beta j$. The quantity $b_\nu^{\alpha i}$, as before, is a function of the $\mathbf{p}_\mu^{\alpha i}$ for all μ; the quantity $C_\nu^{\alpha i,\beta j}$ depends on the $\mathbf{r}_\mu^{\alpha i}$ for all μ and also on the $\mathbf{r}_\mu^{\beta j}$ for all μ; $C_\nu^{\alpha i,\beta j}$ is thus a function of the coordinates for all of the beads belonging to the two molecules αi and βj. Since $\alpha i \neq \beta j$ this quantity is associated with bead-bead interactions between beads belonging to different molecules of the same or different species.

We start by replacing the Dirac delta function in Eq. (5.12) by the integral of the product of two delta functions:

$$\bar{C}_\nu^{\alpha\beta} = \iint \Sigma_i \Sigma_j' b_\nu^{\alpha i} C_\nu^{\alpha i,\beta j} f(x,t) \delta(\mathbf{r}_\nu^{\alpha i} - \mathbf{r}_\nu^{\alpha}) \delta(\mathbf{r}_\nu^{\alpha} - \mathbf{r}) d\mathbf{r}_\nu^{\alpha} dx \tag{5.13}$$

Then we introduce the products of integrals over the delta functions $\delta(\mathbf{r}_\mu^{\alpha i} - \mathbf{r}_\mu^{\alpha})$ for all μ except $\mu \neq \nu$ and $\delta(\mathbf{r}_\mu^{\beta j} - \mathbf{r}_\mu^{\beta})$ for all μ, and $\delta(\mathbf{p}_\mu^{\alpha i} - \mathbf{p}_\mu^{\alpha})$ for all μ, and $\delta(\mathbf{p}_\mu^{\beta j} - \mathbf{p}_\mu^{\beta})$ for all μ. This gives:

$$\bar{C}_\nu^{\alpha\beta} = \iiiint\!\!\int \Sigma_i \Sigma_j' b_\nu^{\alpha i} C_\nu^{\alpha i,\beta j} f(x,t) \delta(\mathbf{r}^{\alpha i} - \mathbf{r}^{\alpha}) \delta(\mathbf{p}^{\alpha i} - \mathbf{p}^{\alpha})$$
$$\cdot \delta(\mathbf{r}^{\beta j} - \mathbf{r}^{\beta}) \delta(\mathbf{p}^{\beta j} - \mathbf{p}^{\beta}) \delta(\mathbf{r}_\nu^{\alpha} - \mathbf{r}) d\mathbf{r}^{\alpha} d\mathbf{p}^{\alpha} d\mathbf{r}^{\beta} d\mathbf{p}^{\beta} dx \tag{5.14}$$

Because of the delta functions, the $b_\nu^{\alpha i}$ can be regarded as functions of the $\mathbf{p}_\mu^{\alpha}$, and hence the index i on $b_\nu^{\alpha i}$ is not needed. The $C_\nu^{\alpha i,\beta j}$ similarly can be regarded as functions of $\mathbf{r}_\nu^{\alpha}$ and $\mathbf{r}_\mu^{\beta}$, and the indices i and j on $C_\nu^{\alpha i,\beta j}$ may now be removed. Next, we perform the integrations over all the x-variables and use the definition for the doublet distribution function given in Eq. (4.6); this gives:

$$\bar{C}_\nu^{\alpha\beta} = \iiiint b_\nu^{\alpha} C_\nu^{\alpha\beta} f_{\alpha\beta}(\mathbf{r}^{\alpha}, \mathbf{p}^{\alpha}, \mathbf{r}^{\beta}, \mathbf{p}^{\beta}, t) \delta(\mathbf{r}_\nu^{\alpha} - \mathbf{r}) d\mathbf{r}^{\alpha} d\mathbf{p}^{\alpha} d\mathbf{r}^{\beta} d\mathbf{p}^{\beta} \tag{5.15}$$

We then integrate over all momenta making use of Eq. (5.2) to get:

$$\bar{C}_\nu^{\alpha\beta} = \iint [[b_\nu^{\alpha}]]^{\alpha\beta} C_\nu^{\alpha\beta} \Psi_{\alpha\beta}(\mathbf{r}^{\alpha}, \mathbf{r}^{\beta}, t) \delta(\mathbf{r}_\nu^{\alpha} - \mathbf{r}) d\mathbf{r}^{\alpha} d\mathbf{r}^{\beta} \tag{5.16}$$

In going from Eq. (5.12) to this expression, all the variables have been integrated out, except for those involving two molecules; the integration over the momenta has been indicated symbolically by use of the double-bracket notation. Note that in Eq. (5.16), when α and β are the same, $\bar{C}_\nu^{\alpha\alpha}$ describes interactions between beads of two different molecules of species α.

We next make the change of variables from $\mathbf{r}_\nu^{\alpha}$ and $\mathbf{r}_\mu^{\beta}$ to $\mathbf{r}_c^{\alpha}, \mathbf{Q}_k^{\alpha}$ and $\mathbf{r}_c^{\beta}$, $\mathbf{Q}_k^{\beta}$:

$$\bar{C}_\nu^{\alpha\beta} = \iiiint [[b_\nu^{\alpha}]]^{\alpha\beta} C_\nu^{\alpha\beta} \Psi_{\alpha\beta}(\mathbf{r}_c^{\alpha}, \mathbf{Q}^{\alpha}, \mathbf{r}_c^{\beta}, \mathbf{Q}^{\beta}, t)$$
$$\cdot \delta(\mathbf{r}_c^{\alpha} + \mathbf{R}_\nu^{\alpha} - \mathbf{r}) d\mathbf{r}_c^{\alpha} d\mathbf{Q}^{\alpha} d\mathbf{r}_c^{\beta} d\mathbf{Q}^{\beta} \tag{5.17}$$

Then we make the further change of variables described in Eqs. (4.10) and (4.11), to obtain:

$$\bar{C}_{\nu}^{\alpha\beta} = \iiiint [[b_{\nu}^{\alpha}]]^{\alpha\beta} C_{\nu}^{\alpha\beta} \tilde{\Psi}_{\alpha\beta}(\mathbf{r}_c^{\alpha}, \mathbf{R}_{\alpha\beta}, \mathbf{Q}^{\alpha}, \mathbf{Q}^{\beta}, t)$$
$$\cdot \delta\left(\mathbf{r}_c^{\alpha\beta} - \frac{m_m^{\beta}}{m_m^{\alpha} + m_m^{\beta}} \mathbf{R}_{\alpha\beta} + \mathbf{R}_{\nu}^{\alpha} - \mathbf{r}\right) d\mathbf{r}_c^{\alpha\beta} d\mathbf{R}_{\alpha\beta} d\mathbf{Q}^{\alpha} d\mathbf{Q}^{\beta} \tag{5.18}$$

Integration over the coordinates of the center of mass $\mathbf{r}_{\nu}^{\alpha\beta}$ of the pair of molecules then gives:

$$\bar{C}_{\nu}^{\alpha\beta} = \iiint [[b_{\nu}^{\alpha}]]^{\alpha\beta} C_{\nu}^{\alpha\beta} \tilde{\Psi}_{\alpha\beta}\left(\mathbf{r} - \mathbf{R}_{\nu}^{\alpha} + \frac{m_m^{\beta}}{m_m^{\alpha} + m_m^{\beta}} \mathbf{R}_{\alpha\beta}, \mathbf{R}_{\alpha\beta}, \mathbf{Q}^{\alpha}, \mathbf{Q}^{\beta}, t\right)$$
$$\cdot d\mathbf{R}_{\alpha\beta} d\mathbf{Q}_{\alpha} d\mathbf{Q}^{\beta} \tag{5.19}$$

Note that the arguments of $[[b_{\nu}^{\alpha}]]^{\alpha\beta}$ and $C_{\nu}^{\alpha\beta}$ are the same as those of the distribution function $\tilde{\Psi}_{\alpha\beta}$. Since we generally assume that $\tilde{\Psi}_{\alpha\beta}$ varies slowly with the first independent variable, the integrand may be expanded in a Taylor series about r, to get:

$$\bar{C}_{\nu}^{\alpha\beta} = \langle \Sigma_i \Sigma_j' b_{\nu}^{\alpha i} C_{\nu}^{\alpha i, \beta j} \delta(\mathbf{r}_{\nu}^{\alpha i} - \mathbf{r}) \rangle$$
$$= \iiint [[b_{\nu}^{\alpha}]]^{\alpha\beta} C_{\nu}^{\alpha\beta} \tilde{\Psi}_{\alpha\beta}(\mathbf{r}, \mathbf{R}_{\alpha\beta}, \mathbf{Q}^{\alpha}, \mathbf{Q}^{\beta}, t) d\mathbf{R}_{\alpha\beta} d\mathbf{Q}^{\alpha} d\mathbf{Q}^{\beta}$$
$$- \nabla \cdot \iiint \left(\mathbf{R}_{\nu}^{\alpha} - \frac{m_m^{\beta}}{m_m^{\alpha} + m_m^{\beta}} \mathbf{R}_{\alpha\beta}\right) [[b_{\nu}^{\alpha}]]^{\alpha\beta}$$
$$\cdot C_{\nu}^{\alpha\beta} \tilde{\Psi}_{\alpha\beta}(\mathbf{r}, \mathbf{R}_{\alpha\beta}, \mathbf{Q}_{\alpha}, \mathbf{Q}^{\beta}, t) d\mathbf{R}_{\alpha\beta} d\mathbf{Q}^{\alpha} d\mathbf{Q}^{\beta} + \cdots \tag{5.20}$$

Here the quantities $[[b_{\nu}^{\alpha}]]^{\alpha\beta}$ and $C_{\nu}^{\alpha\beta}$ depend on $\mathbf{r}$, $\mathbf{R}_{\alpha\beta}$, $\mathbf{Q}^{\alpha}$, $\mathbf{Q}^{\beta}$. Note that, to include all the terms in the expansion formally, the second term in Eq. (5.20) should be replaced by:

$$- \nabla \cdot \iiint \int_0^1 \left(\mathbf{R}_{\nu}^{\alpha} - \frac{m_m^{\beta}}{m_m^{\alpha} + m_m^{\beta}} \mathbf{R}_{\alpha\beta}\right) [[b_{\nu}^{\alpha}]]^{\alpha\beta}$$
$$\cdot C_{\nu}^{\alpha\beta} \tilde{\Psi}_{\alpha\beta}\left(\mathbf{r} - \xi \mathbf{R}_{\nu}^{\alpha} + \xi \frac{m_m^{\beta}}{m_m^{\alpha} + m_m^{\beta}} \mathbf{R}_{\alpha\beta}, \mathbf{R}_{\alpha\beta}, \mathbf{Q}^{\alpha}, \mathbf{Q}^{\beta}, t\right)$$
$$\cdot d\xi d\mathbf{R}_{\alpha\beta} d\mathbf{Q}^{\alpha} d\mathbf{Q}^{\beta} \tag{5.21}$$

Equations (5.19) to (5.21) will be used several times in subsequent developments, where two-molecule interactions are described.

6 The Hydrodynamic Equation of Continuity and the Mass-Flux Vector (DPL, Sect. 17.2a)

By continuum arguments it is known that the equation of continuity for a species α in a multicomponent mixture with no chemical reactions is [Ref. [11], Eq. (18.3–4); Ref. [12], Eq. (11.1–1)]:

$$\frac{\partial}{\partial t}\rho_\alpha = -(\nabla\cdot\rho_\alpha\mathbf{v}_\alpha) = -(\nabla\cdot\rho_\alpha\mathbf{v}) - (\nabla\cdot\mathbf{j}_\alpha) \quad (\alpha = 1, 2, 3 \ldots) \tag{6.1}$$

where ρ_α is the mass concentration of α, $\mathbf{v}_\alpha$ is the velocity of species α, $\mathbf{v}$ is the mass-average velocity of the liquid mixture, and $\mathbf{j}_\alpha$ is the mass flux of α with respect to the mass-average velocity. The mass flux is so defined that $(\mathbf{n}\cdot\mathbf{j}_\alpha)dS$ is the mass of species α moving across a surface element dS per unit time from the negative to the positive side of dS; here $\mathbf{n}$ is the unit vector normal to dS pointing from the negative to the positive side. Frequent use is made of the following relations:

$$\rho(\mathbf{r}, t) = \Sigma_\alpha\rho_\alpha \qquad = \text{mass density of liquid} \tag{6.2}$$

$$\rho(\mathbf{r}, t)\mathbf{v}(\mathbf{r}, t) = \Sigma_\alpha\rho_\alpha\mathbf{v}_\alpha \qquad = \text{momentum density of liquid} \tag{6.3}$$

$$\mathbf{j}_\alpha(\mathbf{r},t) = \rho_\alpha(\mathbf{v}_\alpha - \mathbf{v}) \qquad = \text{mass flux of species } \alpha \tag{6.4}$$

Addition of all the equations in Eq. (6.1) gives $\partial\rho/\partial t = -(\nabla\cdot\rho\mathbf{v})$, the equation of continuity for the liquid mixture, because $\Sigma_\alpha\mathbf{j}_\alpha = \mathbf{0}$.

If we can find a scalar function B_α in the phase space such that $\langle B_\alpha\rangle$ is the mass concentration ρ_α at position $\mathbf{r}$ and time t, then, from the general equation of change, we can generate the equation of continuity and obtain a formal expression for the mass flux vector. The appropriate choice of B_α, which indicates how much mass of α is located in a unit volume surrounding position $\mathbf{r}$, is:

$$B_\alpha = \Sigma_i\Sigma_\nu m_\nu^\alpha\delta(\mathbf{r}_\nu^{\alpha i} - \mathbf{r}) \tag{6.5}$$

so that

$$\rho_\alpha = \langle B_\alpha\rangle = \langle\Sigma_i\Sigma_\nu m_\nu^\alpha\delta(\mathbf{r}_\nu^{\alpha i} - \mathbf{r})\rangle \tag{6.6}$$

Since the Dirac delta function is zero unless bead ν of molecule αi is at $\mathbf{r}$, this expression accounts explicitly for the the extension of the polymer molecules in space. In other words, it is *not* assumed that the mass of a polymer molecule is concentrated at its center of mass, but rather that the mass is distributed at the positions of the beads. This is in sharp contrast with the kinetic theory of polyatomic gases [13], in which the entire mass is presumed to be located at the center of mass of the molecule. The factor $\delta(\mathbf{r}_\nu^{\alpha i} - \mathbf{r})$ included in Eq. (6.5) plays an important role in the ensuing development, as already adumbrated in Eqs. (5.10) and (5.20).

When Eqs. (5.7), (5.9), and (5.10) are successively applied to Eq. (6.6), we get:

$$\begin{aligned}\rho_\alpha &= \Sigma_\nu m_\nu^\alpha \int \bar{\Psi}_\alpha(\mathbf{r}^\alpha, t)\delta(\mathbf{r}_\nu^\alpha - \mathbf{r})d\mathbf{r}^\alpha \\ &= \Sigma_\nu m_\nu^\alpha \int \Psi_\alpha(\mathbf{r} - \mathbf{R}_\nu^\alpha, \mathbf{Q}^\alpha, t)d\mathbf{Q}^\alpha \\ &= \Sigma_\nu m_\nu^\alpha \int \Psi_\alpha(\mathbf{r}, \mathbf{Q}^\alpha, t)d\mathbf{Q}^\alpha - \nabla\cdot\Sigma_\nu m_\nu^\alpha \int \mathbf{R}_\nu^\alpha \Psi_\alpha(\mathbf{r}, \mathbf{Q}^\alpha, t)d\mathbf{Q}^\alpha + \cdots \\ &= n_\alpha m_m^\alpha - 0 + \tfrac{1}{2}\nabla\nabla : \Sigma_\nu m_\nu^\alpha \int \mathbf{R}_\nu^\alpha \mathbf{R}_\nu^\alpha \Psi_\alpha(\mathbf{r}, \mathbf{Q}^\alpha, t)d\mathbf{Q}^\alpha - \cdots \end{aligned} \tag{6.7}$$

in which Eq. (4.5) and $\Sigma_\nu m_\nu^\alpha \mathbf{R}^\alpha = \mathbf{0}$ have been used. Note that the relation $\rho_\alpha = n_\alpha m^\alpha$ is not true if more than two terms are used in the expansion of Eq. (5.10), because of the extension of the polymer molecules in space. This point is discussed further in Appendix B; see also Fig. 4.

When the expression in Eq. (6.5) is inserted into the general equation of change in Eq. (3.7), and when the definition of the Liouville operator in Eq. (3.6) is used, we get:

$$\begin{aligned}\frac{\partial}{\partial t}\rho_\alpha &= \left\langle \sum_{\beta j \mu} \frac{\mathbf{p}_\mu^{\beta j}}{m_\mu^\beta}\cdot\frac{\partial}{\partial \mathbf{r}_\mu^{\beta j}}(\Sigma_i \Sigma_\nu m_\nu^\alpha \delta(\mathbf{r}_\nu^{\alpha i} - \mathbf{r}))\right\rangle \\ &= \left\langle \Sigma_i \Sigma_\nu \frac{\mathbf{p}_\nu^{\alpha i}}{m_\nu^\alpha}\cdot\frac{\partial}{\partial \mathbf{r}_\nu^{\alpha i}}(m_\nu^\alpha \delta(\mathbf{r}_\nu^{\alpha i} - \mathbf{r}))\right\rangle \\ &= -\left\langle \Sigma_i \Sigma_\nu \left(\mathbf{p}_\nu^{\alpha i}\cdot\frac{\partial}{\partial \mathbf{r}}\delta(\mathbf{r}_\nu^{\alpha i} - \mathbf{r})\right)\right\rangle \\ &= -(\nabla\cdot\langle \Sigma_i \Sigma_\nu \mathbf{p}_\nu^{\alpha i}\delta(\mathbf{r}_\nu^{\alpha i} - \mathbf{r})\rangle)\end{aligned} \tag{6.8}$$

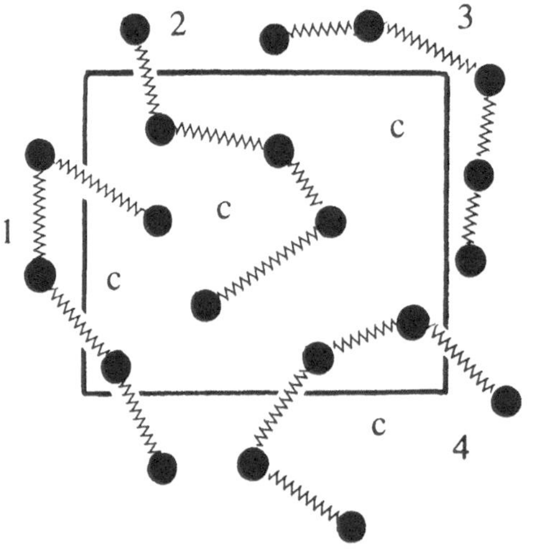

Fig. 4. Sketch showing why ρ_α and n_α are not simply related by $\rho_\alpha = m_m^\alpha n_\alpha$. (a) The mass concentration ρ_α gives the mass of beads (from *any* chains) in a unit volume – here shown as a box – at some point in space , regardless of the locations of the centers of mass "c" of the chains. Note that chain #4 has its center of mass outside the volume element but two of its beads contribute to the mass of beads within the unit volume, and therefore to ρ_α. (b) The number density n_α gives the number of chains whose centers of mass "c" lie within the unit volume. Note that chain #3 contributes to n_α although all its beads are outside the unit volume

In going from the second line to the third line we have used the fact that $(\partial/\partial x)\delta(x-y) = -(\partial/\partial y)\delta(x-y)$; the operators $\partial/\partial \mathbf{r}$ and ∇ are identical, the latter notation being generally used in continuum mechanics.

Equation (6.8) is the equation of continuity for species α, Eq. (6.1), and the $\langle\cdots\rangle$ quantity in the last line is identified[3] as $\rho_\alpha \mathbf{v}_\alpha$, the momentum flux per unit volume associated with species α, since it describes how much momentum of the molecules is located in a unit volume surrounding the point $\mathbf{r}$. Use of Eqs. (5.7), (5.9), and (5.10) allows us to develop $\rho_\alpha \mathbf{v}_\alpha$ as follows:

$$\begin{aligned}
\rho_\alpha \mathbf{v}_\alpha &= \Sigma_\nu \int [[\mathbf{p}_\nu^\alpha]]^\alpha \bar{\Psi}_\alpha(\mathbf{r}^\alpha, t)\delta(\mathbf{r}_\nu^\alpha - \mathbf{r})d\mathbf{r}^\alpha \\
&= \Sigma_\nu \int [[\mathbf{p}_\nu^\alpha]]^\alpha \Psi_\alpha(\mathbf{r} - \mathbf{R}_\nu^\alpha, \mathbf{Q}^\alpha, t)d\mathbf{Q}^\alpha \\
&= \Sigma_\nu \int [[\mathbf{p}_\nu^\alpha]]^\alpha \Psi_\alpha(\mathbf{r}, \mathbf{Q}^\alpha, t)dQ^\alpha \\
&\quad - \nabla \cdot \Sigma_\nu \int \mathbf{R}_\nu^\alpha [[\mathbf{p}_\nu^\alpha]]^\alpha \Psi_\alpha(\mathbf{r}, \mathbf{Q}^\alpha, t)d\mathbf{Q}^\alpha + \cdots
\end{aligned} \tag{6.9}$$

From Eqs. (6.4), (6.7), and (6.9) we find that:

$$\begin{aligned}
\mathbf{j}_\alpha(\mathbf{r}, t) &= \Sigma_\nu \int ([[\mathbf{p}_\nu^\alpha]]^\alpha - m_\nu^\alpha \mathbf{v}(\mathbf{r}, t))\bar{\Psi}_\alpha(\mathbf{r}^\alpha, t)\delta(\mathbf{r}_\nu^\alpha - \mathbf{r})d\mathbf{r}^\alpha \\
&= \Sigma_\nu \int [[\mathbf{p}_\nu^\alpha]]^\alpha \Psi_\alpha(\mathbf{r} - \mathbf{R}_\nu^\alpha, \mathbf{Q}^\alpha, t)d\mathbf{Q}^\alpha \\
&\quad - \mathbf{v}(\mathbf{r}, t)\Sigma_\nu m_\nu^\alpha \int \Psi_\alpha(\mathbf{r} - \mathbf{R}_\nu^\alpha, \mathbf{Q}^\alpha, t)d\mathbf{Q}^\alpha
\end{aligned} \tag{6.10}$$

Note carefully that the arguments of $[[\mathbf{p}_\nu^\alpha]]^\alpha$ are the same as those of Ψ_α, namely $\mathbf{r} - \mathbf{R}_\nu^\alpha$, $\mathbf{Q}^\alpha$, and t. Now for each of the two terms in the second form of Eq. (6.10) we make the Taylor expansion described in Eq. (5.10):

$$\begin{aligned}
\mathbf{j}_\alpha(\mathbf{r}, t) &= \Sigma_\nu m_\nu^\alpha \int [[\dot{\mathbf{r}}_\nu^\alpha]]^\alpha \Psi_\alpha(\mathbf{r}, \mathbf{Q}^\alpha, t)d\mathbf{Q}^\alpha \\
&\quad - \nabla \cdot \Sigma_\nu m_\nu^\alpha \int \mathbf{R}_\nu^\alpha [[\dot{\mathbf{r}}_\nu^\alpha]]^\alpha \Psi_\alpha(\mathbf{r}, \mathbf{Q}^\alpha, t)d\mathbf{Q}^\alpha \\
&\quad + \tfrac{1}{2}\nabla\nabla : \Sigma_\nu m_\nu^\alpha \int \mathbf{R}_\nu^\alpha \mathbf{R}_\nu^\alpha [[\dot{\mathbf{r}}_\nu^\alpha]]^\alpha \Psi_\alpha(\mathbf{r}, \mathbf{Q}^\alpha, t)d\mathbf{Q}^\alpha + \cdots
\end{aligned}$$

---

[3] Since it is only the divergence of $\rho_\nu \mathbf{v}_\alpha$ that appears in Eq. (6.8), one could add to the $\langle\cdots\rangle$ quantity in the last line of Eq. (6.8) any solenoidal vector function. However, it follows from the physical interpretation after Eq. (6.8) that the identification we have made is the correct one. This same point has been discussed by Irving and Kirkwood (see the Appendix to [6]) with regard to the stress tensor. See also Appendix A of this article.

$$- \mathbf{v}(\mathbf{r}, t)\Sigma_\nu m_\nu^\alpha \int \Psi_\alpha(\mathbf{r}, \mathbf{Q}^\alpha, t)d\mathbf{Q}^\alpha$$

$$+ \mathbf{0} - \tfrac{1}{2}\mathbf{v}(\mathbf{r}, t)\nabla\nabla : \Sigma_\nu m_\nu^\alpha \int \mathbf{R}_\nu^\alpha \mathbf{R}_\nu^\alpha \Psi_\alpha(\mathbf{r}, \mathbf{Q}^\alpha, t)d\mathbf{Q}^\alpha + \cdots \tag{6.11}$$

In each series, the first term is the dominant contribution, and it is this zeroth-order contribution that is featured in Table 1 and used in Sect. 15.1. The first-order contribution (containing ∇) is used in Sect. 15.2, and the second-order contribution (containing $\nabla\nabla$) is used in Sect. 15.3. The complete expression in Eq. (6.10) is used in Appendix B.

7 The Hydrodynamic Equation of Motion and the Stress Tensor (DPL, Sect. 17.2b)

From continuum mechanics arguments [11, 12] the equation of motion for a fluid mixture is known to be:[4]

$$\frac{\partial}{\partial t}\rho\mathbf{v} = -[\nabla\cdot\rho\mathbf{v}\mathbf{v}] - [\nabla\cdot\boldsymbol{\pi}] + \mathbf{G} \tag{7.1}$$

where $\boldsymbol{\pi}$ is the stress tensor, and $\mathbf{G}$ is the external force (per unit volume) acting on the fluid. The stress tensor is defined so that $[\mathbf{n}\cdot\boldsymbol{\pi}]\,dS$ is the force transmitted across the element of surface dS from the negative to the positive side of dS.

We now have to find a vector function $\mathbf{B}$ such that $\langle\mathbf{B}\rangle$ is the momentum flux $\rho\mathbf{v}$ at position $\mathbf{r}$ and time t. Because we have already identified the $\langle\cdots\rangle$ expression in the last line of Eq. (6.8) as $\rho_\alpha\mathbf{v}_\alpha$, we take $\mathbf{B}$ here as

$$\mathbf{B} = \sum_{\alpha i\nu} \mathbf{p}_\nu^{\alpha i}\,\delta(\mathbf{r}_\nu^{\alpha i} - \mathbf{r}) \tag{7.2}$$

When this $\mathbf{B}$ is put into the general equation of change, Eq. (3.7), and the steps in Eq. (6.8), are followed, we get:

$$\frac{\partial}{\partial t}\rho\mathbf{v} = -\nabla\cdot\left\langle\sum_{\alpha i\nu}\frac{1}{m_\nu^\alpha}\mathbf{p}_\nu^{\alpha i}\mathbf{p}_\nu^{\alpha i}\delta(\mathbf{r}_\nu^{\alpha i} - \mathbf{r})\right\rangle + \left\langle\sum_{\alpha i\nu}\mathbf{F}_\nu^{\alpha i}\,\delta(\mathbf{r}_\nu^{\alpha i} - \mathbf{r})\right\rangle \tag{7.3}$$

The last term may be regarded as a "momentum source" term $\mathbf{S}$ and can be decomposed into several contributions according to Eq. (2.32)

$$\mathbf{S} = \mathbf{S}^{(\phi)} + \mathbf{S}^{(d)} + \mathbf{S}^{(e)} = \left\langle\sum_{\alpha i\nu}(\mathbf{F}_\nu^{(\phi)\alpha i} + \mathbf{F}_\nu^{(d)\alpha i} + \mathbf{F}_\nu^{(e)\alpha i})\delta(\mathbf{r}_\nu^{\alpha i} - \mathbf{r})\right\rangle \tag{7.4}$$

---

[4] Usually the external force term $\mathbf{G}$ is written as $\Sigma_\alpha\rho_\alpha\mathbf{g}_\alpha$ where $\mathbf{g}_\alpha$ is the acceleration imparted to each chemical species present (see [11], Eq. (18.3-2) and [12], Eq. (11.1-3)). In Eq. 7.11 it turns out that the external force does *not* have this form in a system in which the "beads" of the individual molecules may have different forces acting on them.

In the following subsections we get the various contributions to the stress tensor by applying Eqs. (5.9–5.11), and (5.19–5.21). We ultimately obtain Eq. (7.1), with $\boldsymbol{\pi}$ given as a sum of four contributions: $\boldsymbol{\pi} = \boldsymbol{\pi}^{(k)} + \boldsymbol{\pi}^{(\phi)} + \boldsymbol{\pi}^{(d)} + \boldsymbol{\pi}^{(e)}$, the kinetic, intramolecular, intermolecular, and external contributions.

7.1 The Kinetic Contribution to the Stress Tensor

We start by rewriting the first $\langle \cdots \rangle$ term in Eq. (7.3) by replacing $\mathbf{p}_\nu^\alpha$ by $m_\nu^\alpha(\dot{\mathbf{r}}_\nu^\alpha - \mathbf{v})$, where $\mathbf{v} = \mathbf{v}(\mathbf{r}, t)$, and supplying compensating terms to get

$$\left\langle \sum_{\alpha i \nu} \frac{1}{m_\nu^\alpha} \mathbf{p}_\nu^{\alpha i} \mathbf{p}_\nu^{\alpha i}\, \delta(\mathbf{r}_\nu^{\alpha i} - \mathbf{r}) \right\rangle = \left\langle \sum_{\alpha i \nu} m_\nu^\alpha (\dot{\mathbf{r}}_\nu^{\alpha i} - \mathbf{v})(\dot{\mathbf{r}}_\nu^{\alpha i} - \mathbf{v})\delta(\mathbf{r}_\nu^{\alpha i} - \mathbf{r}) \right\rangle$$
$$+ \left\langle \sum_{\alpha i \nu} m_\nu^\alpha (\dot{\mathbf{r}}_\nu^{\alpha i}\mathbf{v} + \mathbf{v}\dot{\mathbf{r}}_\nu^{\alpha i} - \mathbf{v}\mathbf{v})\delta(\mathbf{r}_\nu^{\alpha i} - \mathbf{r}) \right\rangle \tag{7.5}$$

The second term may be simplified by using the expression for ρ_α in Eq. (6.6) as well as the expression for $\rho_\alpha \mathbf{v}_\alpha$ implied by Eqs. (6.1) and (6.8) to get:

$$\Sigma_\alpha[\rho_\alpha \mathbf{v}_\alpha \mathbf{v} + \mathbf{v}\rho_\alpha \mathbf{v}_\alpha - \rho_\alpha \mathbf{v}\mathbf{v}] = \rho\mathbf{v}\mathbf{v} \tag{7.6}$$

Equation (6.3) for $\rho\mathbf{v}$ was also used. Hence Eq. (7.3) may now be written as:

$$\frac{\partial}{\partial t}\rho\mathbf{v} = -[\nabla\cdot\rho\mathbf{v}\mathbf{v}] - [\nabla\cdot\boldsymbol{\pi}^{(k)}] + \mathbf{S} \tag{7.7}$$

in which $\boldsymbol{\pi}^{(k)}$, the kinetic contribution to the stress tensor, is the first term on the right side of Eq. (7.5). This quantity, which is the momentum flux resulting from the motion of the beads across a surface moving with velocity $\mathbf{v}(\mathbf{r}, t)$, can be put in the form of Eq. (5.9) or Eq. (5.11).

$$\boldsymbol{\pi}^{(k)} = \sum_{\alpha\nu} \int m_\nu^\alpha [[(\dot{\mathbf{r}}_\nu^\alpha - \mathbf{v})(\dot{\mathbf{r}}_\nu^\alpha - \mathbf{v})]]^\alpha \Psi_\alpha(\mathbf{r} - \mathbf{R}_\nu^\alpha, \mathbf{Q}^\alpha, t)d\mathbf{Q}^\alpha$$
$$= \sum_{\alpha\nu} m_\nu^\alpha \int [[(\dot{\mathbf{r}}_\nu^\alpha - \mathbf{v})(\dot{\mathbf{r}}_\nu^\alpha - \mathbf{v})]]^\alpha \Psi_\alpha(\mathbf{r}, \mathbf{Q}^\alpha, t)d\mathbf{Q}^\alpha + \cdots \tag{7.8}$$

Note, however, that before writing out the higher terms in the expansion in second line of Eq. (7.8) by applying Eq. (5.11), the $\mathbf{v}(\mathbf{r}, t)$ must be taken *outside* the integral so that it will *not* be operated on by the spatial derivative.

When the product in the double-bracket in the first line of Eq. (7.8) is multiplied out, the resulting $[[\dot{\mathbf{r}}_\nu^\alpha \dot{\mathbf{r}}_\nu^\alpha]]^\alpha$ and $[[\dot{\mathbf{r}}_\nu^\alpha]]^\alpha$ are functions of $\mathbf{r} - \mathbf{R}_\nu^\alpha$, $\mathbf{Q}^\alpha$, and t. The velocity $\mathbf{v}$, on the other hand, is a function of $\mathbf{r}$ and t. Tables 1 and 2 are summaries of the various contributions to the flux expressions; Table 1 shows the first term in the Taylor-series expansion in the fluxes, whereas Table 2 gives the complete expressions. Thus, for the kinetic contribution to the stress tensor, in Table 1 we find the first term of the series in the second line of Eq. (7.8), whereas in Table 2 we find the complete expression – containing the displaced coordinate – from which the Taylor series expansion can be generated if desired.

7.2 The External Force Contribution to the Stress Tensor

From Eqs. (7.3 and 7.4) we get for the external force contribution:

$$\mathbf{S}^{(e)} = \left\langle \sum_{\alpha i \nu} \mathbf{F}_{\nu}^{(e)\alpha i}\, \delta(\mathbf{r}_{\nu}^{\alpha i} - \mathbf{r}) \right\rangle \tag{7.9}$$

Applying Eqs. (5.9) and (5.11) to this phase-space average gives:

$$\begin{aligned} \mathbf{S}^{(e)} &= \Sigma_\alpha \Sigma_\nu \int \mathbf{F}_{\nu}^{(e)\alpha} \Psi_\alpha(\mathbf{r} - \mathbf{R}_\nu^\alpha, \mathbf{Q}^\alpha, t)\, d\mathbf{Q}^\alpha \\ &= \Sigma_\alpha \int \mathbf{F}^{(e)\alpha} \Psi_\alpha(\mathbf{r}, \mathbf{Q}^\alpha, t)\, d\mathbf{Q}^\alpha \\ &\quad - \left[\nabla \cdot \Sigma_\alpha \Sigma_\nu \int \int_0^1 \mathbf{R}_\nu^\alpha \mathbf{F}_\nu^{(e)\alpha} \Psi_\alpha(\mathbf{r} - \xi \mathbf{R}_\nu^\alpha, \mathbf{Q}^\alpha, t)\, d\xi\, d\mathbf{Q}^\alpha \right] \end{aligned} \tag{7.10}$$

where $\mathbf{F}^{(e)\alpha}$ is defined in Eq. (2.34). Comparison with Eq. (7.1) allows us to make the identifications:

$$\mathbf{G} = \Sigma_\alpha \int \mathbf{F}^{(e)\alpha}(\mathbf{r}, \mathbf{Q}^\alpha, t)\, d\mathbf{Q}^\alpha \tag{7.11}$$

$$\begin{aligned} \boldsymbol{\pi}^{(e)} &= \Sigma_\alpha \Sigma_\nu \int \int_0^1 \mathbf{R}_\nu^\alpha \mathbf{F}_\nu^{(e)\alpha} \Psi_\alpha(\mathbf{r} - \xi \mathbf{R}_\nu^\alpha, \mathbf{Q}^\alpha, t)\, d\xi\, d\mathbf{Q}^\alpha \\ &= \Sigma_\alpha \Sigma_\nu \int \mathbf{R}_\nu^\alpha \mathbf{F}_\nu^{(e)\alpha} \Psi_\alpha(\mathbf{r}, \mathbf{Q}^\alpha, t)\, d\mathbf{Q}^\alpha + \cdots \end{aligned} \tag{7.12}$$

The first line of Eq. (7.12) is the complete external-force contribution to the stress tensor given in Table 2, and the first term in the series in the second line is given in Table 1. If the external forces per unit mass are the same for all beads of a given molecule, so that all $(\mathbf{F}_\nu^{(e)\alpha}/m_\nu^\alpha)$ are constant and equal to $\mathbf{g}_\alpha$, then[4] $\mathbf{G} = \Sigma_\alpha \rho_\alpha \mathbf{g}_\alpha$, and $\boldsymbol{\pi}^{(e)}$ is equal to zero (since $\Sigma_\nu m_\nu^\alpha \mathbf{R}_\nu^\alpha = 0$). A result analogous to that in Eq. (7.12) does not appear in Irving and Kirkwood [6], since they considered only monatomic liquids with identical atoms. For a further discussions of the external force contribution to the stress tensor, see Appendix A and also Sect. 14.4.

7.3 The Intramolecular Contribution to the Stress Tensor

The intramolecular contribution to the source term is, from Eq. (7.4):

$$\mathbf{S}^{(\phi)} = \left\langle \sum_{\alpha i \nu} \mathbf{F}_{\nu}^{(\phi)\alpha i}\, \delta(\mathbf{r}_{\nu}^{\alpha} - \mathbf{r}) \right\rangle \tag{7.13}$$

Application of Eqs. (5.9) and (5.11) then gives:

$$\mathbf{S}^{(\phi)} = \Sigma_\alpha \Sigma_\nu \int \mathbf{F}_\nu^{(\phi)\alpha} \Psi_\alpha(\mathbf{r} - \mathbf{R}_\nu^\alpha, \mathbf{Q}^\alpha, t) d\mathbf{Q}^\alpha$$

$$= \Sigma_\alpha \Sigma_\nu \int \mathbf{F}_\nu^{(\phi)\alpha} \Psi_\alpha(\mathbf{r}, \mathbf{Q}^\alpha, t) d\mathbf{Q}^\alpha - \left[\nabla \cdot \Sigma_\alpha \Sigma_\nu \int \int_0^1 \mathbf{R}_\nu^\alpha \mathbf{F}_\nu^{(\phi)\alpha} \right.$$

$$\left. \cdot \Psi_\alpha(\mathbf{r} - \xi \mathbf{R}_\nu^\alpha, \mathbf{Q}^\alpha, t) \, d\xi d\mathbf{Q}^\alpha \right] \tag{7.14}$$

The first term vanishes since $\Sigma_\nu \mathbf{F}_\nu^{(\phi)\alpha} = \Sigma_\nu \Sigma_\mu \mathbf{F}_{\nu\mu}^{(\phi)\alpha} = 0$. The second term can be identified as $-[\nabla \cdot \boldsymbol{\pi}^{(\phi)}]$. Therefore

$$\boldsymbol{\pi}^{(\phi)} = \Sigma_\alpha \Sigma_\nu \int \int_0^1 \mathbf{R}_\nu^\alpha \mathbf{F}_\nu^{(\phi)\alpha} \Psi_\alpha(\mathbf{r} - \xi \mathbf{R}_\nu^\alpha, \mathbf{Q}^\alpha, t) \, d\xi d\mathbf{Q}^\alpha$$

$$= \tfrac{1}{2} \sum_{\alpha\nu\mu} \int \mathbf{R}_{\nu\mu}^\alpha \mathbf{F}_{\nu\mu}^{(\phi)\alpha} \Psi_\alpha(\mathbf{r}, \mathbf{Q}^\alpha, t) d\mathbf{Q}^\alpha + \cdots \tag{7.15}$$

which is the contribution of the "springs" (and excluded-volume forces) in single molecules to the stress tensor, i.e. the intramolecular force contribution. The first line of Eq. (7.15) is given in Table 2, and the first term of the second line in Table 1.

7.4 *The Intermolecular Force Contribution to the Stress Tensor*

The quantity $\mathbf{S}^{(d)}$ is the intermolecular force contribution to the source in Eq. (7.4). It accounts for the forces between beads on different molecules, and is given by:

$$\mathbf{S}^{(d)} = \left\langle \sum_{\alpha i \nu} \mathbf{F}_\nu^{(d)\alpha i} \delta(\mathbf{r}_\nu^{\alpha i} - \mathbf{r}) \right\rangle = \left\langle \sum_{\alpha i \nu} \sum_{\beta j \mu} \mathbf{F}_{\nu\mu}^{(d)\alpha i, \beta j} \delta(\mathbf{r}_\nu^{\alpha i} - \mathbf{r}) \right\rangle \tag{7.16}$$

This is of the form of Eq. (5.12), and therefore, Eq. (5.20) can be applied. The first term in the expansion in Eq. (5.20) gives zero because of $\mathbf{F}_{\nu\mu}^{\alpha\beta}(R_{\nu\mu}^{\alpha\beta}) = -\mathbf{F}_{\mu\nu}^{\beta\alpha}(R_{\mu\nu}^{\beta\alpha})$, which is a statement of Newton's third law for pairs of beads; the interbead distance is the absolute value of $\mathbf{R}_{\nu\mu}^{\alpha\beta} = \mathbf{r}_\mu^\beta - \mathbf{r}_\nu^\alpha$ (cf. Eq. 2.21). The second term gives minus the gradient of the intermolecular contribution to the stress tensor, which is:

$$\boldsymbol{\pi}^{(d)} = \sum_{\alpha\beta\nu\mu} \int\int\int \left(\mathbf{R}_\nu^\alpha - \frac{m_m^\beta}{m_m^\alpha + m_m^\beta} \mathbf{R}_{\alpha\beta} \right) \mathbf{F}_{\nu\mu}^{(d)\alpha\beta} \tilde{\Psi}_{\alpha\beta}(\mathbf{r}, \mathbf{R}_{\alpha\beta}, \mathbf{Q}^\alpha, \mathbf{Q}^\beta, t)$$

$$\cdot d\mathbf{R}_{\alpha\beta} d\mathbf{Q}^\alpha d\mathbf{Q}^\beta \tag{7.17}$$

If higher terms in the Taylor-series development are desired, then Eq. (5.21) can be used. Now making use of the symmetry property of $\tilde{\Psi}_{\alpha\beta}$ in Eq. (4.13) and again using Newton's third law, we can rewrite the second term (containing $\mathbf{R}_{\alpha\beta}$)

of Eq. (7.17); this gives

$$\boldsymbol{\pi}^{(d)} = \sum_{\alpha\beta\nu\mu} \iiint (\mathbf{R}_\nu^\alpha - \tfrac{1}{2}\mathbf{R}_{\alpha\beta})\, \mathbf{F}_{\nu\mu}^{(d)\alpha\beta}\, \tilde{\Psi}_{\alpha\beta}(\mathbf{r}, \mathbf{R}_{\alpha\beta}, \mathbf{Q}^\alpha, \mathbf{Q}^\beta, t)\, d\mathbf{R}_{\alpha\beta} d\mathbf{Q}^\alpha d\mathbf{Q}^\beta \tag{7.18}$$

Equation (7.18) can be rewritten with all $\alpha\nu$ and $\beta\mu$ indices interchanged. Then the symmetry of the distribution function (see (Eq. 4.13)) may be used, as well as the application of Newton's third law to pairs of beads. The new expression can then be added to Eq. (7.18) and divided by 2 to give:

$$\boldsymbol{\pi}^{(d)} = \tfrac{1}{2} \sum_{\alpha\beta\nu\mu} \iiint \mathbf{R}_{\mu\nu}^{\beta\alpha} \mathbf{F}_{\nu\mu}^{(d)\alpha\beta}\, \tilde{\Psi}_{\alpha\beta}(\mathbf{r}, \mathbf{R}_{\alpha\beta}, \mathbf{Q}^\alpha, \mathbf{Q}^\beta, t)\, d\mathbf{R}_{\alpha\beta} d\mathbf{Q}^\alpha d\mathbf{Q}^\beta \tag{7.19}$$

In obtaining this result we used the fact that $\mathbf{R}_\nu^\alpha - \mathbf{R}_{\alpha\beta} - \mathbf{R}_\mu^\beta = \mathbf{R}_{\mu\nu}^{\beta\alpha}$ (cf. Eqs (2.21) and (2.22)).

This completes the discussion of the various contributions to the stress tensor. Table 1 gives a summary of the expressions for the flux expressions that are obtained by taking the first term in the Taylor-series expansions of the fluxes, and Table 2 summarizes the complete expressions (except for the intermolecular contributions).

8 The Hydrodynamic Energy Equation and the Heat-Flux Vector (see Ref. [6])

From continuum mechanics we know that the energy equation for a multicomponent mixture has the form [11, Eq. B in Table 18.3-1]:

$$\frac{\partial}{\partial t}(\tfrac{1}{2}\rho v^2 + \rho\hat{U}) = -(\nabla\cdot(\tfrac{1}{2}\rho v^2 + \rho\hat{U})\mathbf{v}) - (\nabla\cdot\mathbf{q}) - (\nabla\cdot[\boldsymbol{\pi}\cdot\mathbf{v}]) + (\mathbf{v}\cdot\mathbf{G}) + J \qquad (8.1)^5$$

in which the last three terms represent, respectively, the work done on an element of the fluid by the stresses, external forces, and diffusional effects. The quantity $\mathbf{q}$ is the heat flux vector, and $\hat{U}$ is the internal energy per unit mass

[5] By subtracting from Eq. (8.1) the scalar product of $\mathbf{v}$ with the equation of motion in Eq. (7.1), one obtains the equation of change for the internal energy:

$$\frac{\partial}{\partial t}\rho\hat{U} = -(\nabla\cdot\rho\hat{U}\mathbf{v}) - (\nabla\cdot\mathbf{q}) - (\boldsymbol{\pi}^\dagger : \nabla\mathbf{v}) + J \tag{8.1a}$$

in which $\boldsymbol{\pi}^\dagger$ is the transpose of the stress tensor. An equation of change for the temperature is developed in Appendix C.

(excluding the kinetic energy associated with the flow velocity $\mathbf{v}$, and excluding any contribution associated with external forces).

We must now find a scalar function B, such that $\langle B \rangle$ gives the energy per unit volume $\frac{1}{2}\rho v^2 + \rho\hat{U}$. This function is [6]:

$$B = \sum_{\alpha i \nu} \left(\frac{1}{2m_\nu^\alpha} \mathbf{p}_\nu^{\alpha i} \cdot \mathbf{p}_\nu^{\alpha i} + U_\nu^{\alpha i} \right) \delta(\mathbf{r}_\nu^{\alpha i} - \mathbf{r}) \tag{8.2}$$

in which $U_\nu^{\alpha i}$ represents the potential energy associated with each of the beads, resulting from both intra- and intermolecular bead-bead interactions; it is a function of the $\mathbf{r}_\mu^{\beta j}$ for all values of β, j, and μ. We follow Irving and Kirkwood [6, Eq. (3.7)] and assign one half of the bead-bead potential energy of interaction to each of the relevant beads at their respective positions. That is, each bead of a pair is assigned its "fair share" of the energy of interaction for the pair of beads. This means that $U_\nu^{\alpha i}$ has the form:

$$U_\nu^{\alpha i} = \tfrac{1}{2} \sum_{\beta j \mu} \phi_{\nu\mu}^{\alpha i, \beta j} = \tfrac{1}{2} \sum_{\mu} \phi_{\nu\mu}^{\alpha i} + \tfrac{1}{2} \sum_{\beta j \mu} \phi_{\nu\mu}^{(d)\alpha i, \beta j} \tag{8.3}$$

The average value of B is then given by:

$$\langle B \rangle = \left\langle \sum_{\alpha i \nu} \left(\tfrac{1}{2} m_\nu^\alpha v^2 + \tfrac{1}{2} m_\nu^\alpha \left(\frac{\mathbf{p}_\nu^{\alpha i}}{m_\nu^\alpha} - \mathbf{v} \right)^2 + U_\nu^{\alpha i} \right) \delta(\mathbf{r}_\nu^{\alpha i} - \mathbf{r}) \right\rangle \tag{8.4}$$

It is now apparent that $\langle B \rangle = \frac{1}{2}\rho v^2 + \rho\hat{U}$: the terms $\frac{1}{2} m_\nu^\alpha v^2$, when summed and averaged, give $\frac{1}{2}\rho v^2$, the kinetic energy associated with the mean motion; the remaining terms, when summed and averaged, give $\rho\hat{U}$, which includes the kinetic energies of the beads relative to the flow velocity $\mathbf{v}$ plus the potential energies, excluding the potential energies associated with the external forces.

When B is inserted into the general equation of change Eq. (3.7), there results:

$$\begin{aligned}
\frac{\partial}{\partial t}(\tfrac{1}{2}\rho v^2 + \rho\hat{U}) = &-\nabla \cdot \left\langle \sum_{\alpha i \nu} \dot{\mathbf{r}}_\nu^{\alpha i} (\tfrac{1}{2} m_\nu^\alpha (\dot{\mathbf{r}}_\nu^{\alpha i} \cdot \dot{\mathbf{r}}_\nu^{\alpha i}) + U_\nu^{\alpha i}) \delta(\mathbf{r}_\nu^{\alpha i} - \mathbf{r}) \right\rangle \\
&+ \left\langle \sum_{\alpha i \nu} \sum_{\beta j \mu} \left(\dot{\mathbf{r}}_\nu^{\alpha i} \cdot \frac{\partial}{\partial \mathbf{r}_\nu^{\alpha i}} U_\mu^{\beta j} \right) \delta(\mathbf{r}_\mu^{\beta j} - \mathbf{r}) \right\rangle \\
&+ \left\langle \sum_{\alpha i \nu} (\dot{\mathbf{r}}_\nu^{\alpha i} \cdot \mathbf{F}_\nu^{\alpha i}) \delta(\mathbf{r}_\nu^{\alpha i} - \mathbf{r}) \right\rangle
\end{aligned} \tag{8.5}$$

In order to put this in the form of Eq. (8.1) we begin by replacing $\dot{\mathbf{r}}_\nu^{\alpha i}$, just after the summation sign in the first line, by $(\dot{\mathbf{r}}_\nu^{\alpha i} - \mathbf{v})$, where $\mathbf{v}(\mathbf{r}, t)$ is the mass-average velocity of the fluid. Of course, a compensating term has to be added, the latter then giving the first term on the right side of the following equation:

$$\frac{\partial}{\partial t}(\tfrac{1}{2}\rho v^2 + \rho\hat{U}) = -(\nabla \cdot (\tfrac{1}{2}\rho v^2 + \rho\hat{U})\mathbf{v}) + Q^{(k)} + Q^{(e)} + Q^{(\Phi)} \tag{8.6}$$

Here, by virtue of Eq. (2.32), the "source terms" are:

$$Q^{(k)} = -\nabla\cdot\left\langle \sum_{\alpha i\nu} (\dot{\mathbf{r}}_\nu^{\alpha i} - \mathbf{v})(\tfrac{1}{2} m_\nu^\alpha \dot{\mathbf{r}}_\nu^{\alpha i}\cdot\dot{\mathbf{r}}_\nu^{\alpha i} + U_\nu^{\alpha i})\delta(\mathbf{r}_\nu^{\alpha i} - \mathbf{r})\right\rangle \tag{8.7}$$

$$Q^{(\Phi)} = \left\langle \sum_{\alpha i\nu} (\dot{\mathbf{r}}_\nu^{\alpha i}\cdot\mathbf{F}_\nu^{(\Phi)\alpha i})\delta(\mathbf{r}_\nu^{\alpha i} - \mathbf{r})\right\rangle$$

$$+\tfrac{1}{2}\left\langle \sum_{\alpha i\nu}\sum_{\beta j\mu}\left(\dot{\mathbf{r}}_\nu^{\alpha i}\cdot\frac{\partial}{\partial\mathbf{r}_\nu^{\alpha i}}\sum_{\gamma k\eta}\phi_{\mu\eta}^{\beta j,\gamma k}\right)\delta(\mathbf{r}_\mu^{\beta j} - \mathbf{r})\right\rangle \tag{8.8}$$

$$Q^{(e)} = \left\langle \sum_{\alpha i\nu} (\dot{\mathbf{r}}_\nu^{\alpha i}\cdot\mathbf{F}_\nu^{(e)\alpha i})\,\delta(\mathbf{r}_\nu^{\alpha i} - \mathbf{r})\right\rangle \tag{8.9}$$

We now manipulate each of the "source terms" by using Eqs. (5.9–5.11) and (5.19–21) in order to obtain Eq. (8.1), with the heat-flux vector being the sum of four contributions: $\mathbf{q} = \mathbf{q}^{(k)} + \mathbf{q}^{(\phi)} + \mathbf{q}^{(d)} + \mathbf{q}^{(e)}$.

8.1 The Kinetic Contribution to the Heat Flux Vector

First, in the kinetic contribution $Q^{(k)}$ we replace both of the $\dot{\mathbf{r}}_\nu^{\alpha i}$ in the dot product by $\dot{\mathbf{r}}_\nu^{\alpha i} - \mathbf{v}$ and put in compensating terms; the mass-average velocity $\mathbf{v}$ is a function of $\mathbf{r}$ and t. This gives:

$$Q^{(k)} = -\nabla\cdot\left\langle \sum_{\alpha i\nu} (\dot{\mathbf{r}}_\nu^{\alpha i} - \mathbf{v})(\tfrac{1}{2} m_\nu^\alpha (\dot{\mathbf{r}}_\nu^{\alpha i} - \mathbf{v})^2 + U_\nu^{\alpha i})\delta(\mathbf{r}_\nu^{\alpha i} - \mathbf{r})\right\rangle$$

$$-\nabla\cdot\left\langle \sum_{\alpha i\nu} m_\nu^\alpha (\dot{\mathbf{r}}_\nu^{\alpha i} - \mathbf{v})(\dot{\mathbf{r}}_\nu^{\alpha i} - \mathbf{v})\cdot\mathbf{v}\delta(\mathbf{r}_\nu^{\alpha i} - \mathbf{r})\right\rangle$$

$$= -(\nabla\cdot\mathbf{q}^{(k)}) - (\nabla\cdot[\boldsymbol{\pi}^{(k)}\cdot\mathbf{v}]) \tag{8.10}$$

which serves to define $\mathbf{q}^{(k)}$, the kinetic contribution to the heat-flux vector; the kinetic contribution to the stress tensor $\boldsymbol{\pi}^{(k)}$ was defined in Eq. (7.8). We now apply Eqs. (5.9) and (5.19) to the $\langle \cdots \rangle$ expression in the first line of Eq. (8.10) and use Eq. (8.3); this gives:

$$\mathbf{q}^{(k)} = \sum_{\alpha\nu} \tfrac{1}{2} m_\nu^\alpha \int [[(\dot{\mathbf{r}}_\nu^\alpha - \mathbf{v})(\dot{\mathbf{r}}_\nu^\alpha - \mathbf{v})^2]]^\alpha \Psi_\alpha(\mathbf{r} - \mathbf{R}_\nu^\alpha, \mathbf{Q}^\alpha, t)\, d\mathbf{Q}^\alpha$$

$$+\tfrac{1}{2}\sum_{\alpha\nu\mu}\int \phi_{\nu\mu}^\alpha [[\dot{\mathbf{r}}_\nu^\alpha - \mathbf{v}]]^\alpha \Psi_\alpha(\mathbf{r} - \mathbf{R}_\nu^\alpha, \mathbf{Q}^\alpha, t) d\mathbf{Q}^\alpha$$

$$+\tfrac{1}{2}\sum_{\alpha\beta\nu\mu}\iiint \phi_{\nu\mu}^{(d)\alpha\beta}[[\dot{\mathbf{r}}_\nu^\alpha - \mathbf{v}]]^{\alpha\beta}$$

$$\cdot\tilde{\Psi}_{\alpha\beta}\left(\mathbf{r} - \mathbf{R}_\nu^\alpha + \frac{m_m^\beta}{m_m^\alpha + m_m^\beta}, \mathbf{R}_{\alpha\beta}, \mathbf{Q}^\alpha, \mathbf{Q}^\beta, t\right) d\mathbf{R}_{\alpha\beta} d\mathbf{Q}^\alpha d\mathbf{Q}^\beta \tag{8.11}$$

This expression is included in Table 2; the first term of the Taylor series expansion of Eq. (8.11) is given in Table 1. The arguments of the square-bracket quantities are those of the distribution function that follows, but $\mathbf{v}$ is a function of $\mathbf{r}$ and t.

The first term in $\mathbf{q}^{(k)}$, when dotted into $\mathbf{n}$ and multiplied by dS, describes the rate of diffusion of kinetic energy across an element of surface dS moving with velocity $\mathbf{v}(\mathbf{r}, t)$. This term corresponds to the expression that arises in the kinetic theory of dilute monatomic gases, where the only mechanism for energy transport is the flux of kinetic energy. It should be noted that in this expression the normal unit vector $\mathbf{n}$ is dotted into $(\dot{\mathbf{r}}_\nu^{\alpha i} - \mathbf{v})$, indicating that diffusion at a small angle to the surface is less effective for transporting kinetic energy than diffusion at a large angle. The second and third terms, when dotted into $\mathbf{n}$ and multiplied by dS, describe the rate of diffusion of intra- and intermolecular potential energy across dS. That is, as the beads move relative to $\mathbf{v}$, they carry with them their "fair share" of the potential energy of interaction.

8.2 The External Force Contribution to the Heat Flux Vector

Next we treat the source term associated with external forces. We first put Eq. (8.9) in the form of Eqs. (5.9) and (5.11):

$$\begin{aligned} Q^{(e)} &= \sum_{\alpha\nu} \int [[\dot{\mathbf{r}}_\nu^\alpha]]^\alpha \cdot \mathbf{F}_\nu^{(e)\alpha} \Psi_\alpha(\mathbf{r} - \mathbf{R}_\nu^\alpha, \mathbf{Q}^\alpha, t) d\mathbf{Q}^\alpha \\ &= \sum_{\alpha\nu} \int [[\dot{\mathbf{r}}_\nu^\alpha]]^\alpha \cdot \mathbf{F}_\nu^{(e)\alpha} \Psi_\alpha(\mathbf{r}, \mathbf{Q}^\alpha, t) d\mathbf{Q}^\alpha \\ &\quad - \left[\nabla \cdot \sum_{\alpha\nu} \int \int_0^1 \mathbf{R}_\nu^\alpha \mathbf{F}_\nu^{(e)\alpha} \cdot [[\dot{\mathbf{r}}_\nu^{\alpha i}]]^\alpha \Psi_\alpha(\mathbf{r} - \xi \mathbf{R}_\nu^\alpha, \mathbf{Q}^\alpha, t) d\xi d\mathbf{Q}^\alpha \right] \end{aligned} \tag{8.12}$$

We now replace $\dot{\mathbf{r}}_\nu^{\alpha i}$ by $(\dot{\mathbf{r}}_\nu^{\alpha i} - \mathbf{v})$, where $\mathbf{v} = \mathbf{v}(\mathbf{r}, t)$ and add compensating terms:

$$\begin{aligned} Q^{(e)} &= \sum_{\alpha\nu} \int [[\dot{\mathbf{r}}_\nu^\alpha - \mathbf{v}(\mathbf{r}, t)]]^\alpha \cdot \mathbf{F}_\nu^{(e)\alpha} \Psi_\alpha(\mathbf{r}, \mathbf{Q}^\alpha, t) d\mathbf{Q}^\alpha \\ &\quad + \mathbf{v}(\mathbf{r}, t) \cdot \sum_{\alpha\nu} \int \mathbf{F}_\nu^{(e)\alpha} \Psi_\alpha(\mathbf{r}, \mathbf{Q}^\alpha, t) d\mathbf{Q}^\alpha \\ &\quad - \nabla \cdot \sum_{\alpha\nu} \int \int_0^1 \mathbf{R}_\nu^\alpha \mathbf{F}_\nu^{(e)\alpha} \cdot [[\dot{\mathbf{r}}_\nu^\alpha - \mathbf{v}(\mathbf{r}, t)]]^\alpha \Psi_\alpha(\mathbf{r} - \xi \mathbf{R}_\nu^\alpha, \mathbf{Q}^\alpha, t) d\xi d\mathbf{Q}^\alpha \\ &\quad - \nabla \cdot \left[\sum_{\alpha\nu} \int \int_0^1 \mathbf{R}_\nu^\alpha \mathbf{F}_\nu^{(e)\alpha} \Psi_\alpha(\mathbf{r} - \xi \mathbf{R}_\nu^\alpha, \mathbf{Q}^\alpha, t) d\xi d\mathbf{Q}^\alpha \cdot \mathbf{v}(\mathbf{r}, t) \right] \\ &= J + (\mathbf{v} \cdot \mathbf{G}) - (\nabla \cdot \mathbf{q}^{(e)}) - (\nabla \cdot [\boldsymbol{\pi}^{(e)} \cdot \mathbf{v}]) \end{aligned} \tag{8.13}$$

The second and fourth terms are identified with the help of Eqs. (7.11) and (7.12). The third term is the divergence of the external force contribution to the heat-flux vector, which is:

$$\mathbf{q}^{(e)} = \sum_{\alpha\nu} \int\int_0^1 \mathbf{R}_\nu^\alpha \mathbf{F}_\nu^{(e)\alpha} \cdot [[\dot{\mathbf{r}}_\nu^\alpha - \mathbf{v}]]^\alpha \Psi_\alpha(\mathbf{r} - \xi\mathbf{R}_\nu^\alpha, \mathbf{Q}^\alpha, t) d\xi d\mathbf{Q}^\alpha$$

$$= \sum_{\alpha\nu} \int \mathbf{R}_\nu^\alpha \mathbf{F}_\nu^{(e)\alpha} \cdot [[\dot{\mathbf{r}}_\nu^\alpha - \mathbf{v}]]^\alpha \Psi_\alpha(\mathbf{r}, \mathbf{Q}^\alpha, t) d\mathbf{Q}^\alpha + \cdots \tag{8.14}$$

The first line of Eq. (8.14) is included in Table 2, and the first term of the second line in Table 1.

The term $\mathbf{q}^{(e)}$, when dotted into $\mathbf{n}$ and multiplied by dS, is the rate of work that is done as the beads diffuse under the external force that is acting on the beads. Note that in this term, the unit normal vector is dotted into the position vector rather than into the relative velocity vector. This means that the external forces acting on various beads are weighted differently because of the extension in space of the molecule.

The first term in Eq. (8.13) is the "diffusional source term":

$$J = \sum_{\alpha\nu} \int [[\dot{\mathbf{r}}_\nu^\alpha - \mathbf{v}]]^\alpha \cdot \mathbf{F}_\nu^{(e)\alpha} \Psi_\alpha(\mathbf{r}, \mathbf{Q}^\alpha, t) d\mathbf{Q}^\alpha \tag{8.15}$$

If all the quantities $\mathbf{F}_\nu^{(e)\alpha}/m_\nu^\alpha = \mathbf{g}_\alpha$ are constant and independent of ν, then, by the first line of Eq. (6.11), the term J becomes $\Sigma_\alpha(\mathbf{j}_\alpha \cdot \mathbf{g}_\alpha)$, and it is this expression that is generally seen in continuum mechanics treatments (see, e.g., Eq. D of Table 18.3-1 of Ref. [11]).

8.3 The Intramolecular Contribution to the Heat-Flux Vector

Next we turn to the source term $Q^{(\Phi)}$ given in Eq. (8.8), which accounts for all intra- and intermolecular forces; with the help of Eq. (8.3) this can be written as follows:[6]

$$Q^{(\Phi)} = + \left\langle \sum_{\alpha i\nu} \sum_{\beta j\mu} (\dot{\mathbf{r}}_\nu^{\alpha i} \cdot \mathbf{F}_{\nu\mu}^{(\Phi)\alpha i, \beta j}) \delta(\mathbf{r}_\nu^{\alpha i} - \mathbf{r}) \right\rangle$$

from 1st $\langle\cdots\rangle$ term in Eq. (8.8)

$$- \tfrac{1}{2} \left\langle \sum_{\alpha i\nu} \sum_{\gamma k\eta} (\dot{\mathbf{r}}_\nu^{\alpha i} \cdot \mathbf{F}_{\nu\eta}^{(\Phi)\alpha i, \gamma k}) \delta(\mathbf{r}_\nu^{\alpha i} - \mathbf{r}) \right\rangle$$

from 2nd $\langle\cdots\rangle$ term in Eq. (8.8)

[6] The last line of Eq. 8.16 can also be written in such a way that the difference of two delta functions appears: $\delta(\mathbf{r}_\nu^{\alpha i} - \mathbf{r}) - \delta(\mathbf{r}_\mu^{\beta j} - \mathbf{r})$. This quantity can then be expanded in a Taylor series, according to the procedure used by Irving and Kirkwood [6]. This method also yields Eqs. (8.18) and (8.21).

$$-\tfrac{1}{2}\left\langle \sum_{\alpha i \nu}\sum_{\beta j \mu} (\dot{\mathbf{r}}_\nu^{\alpha i}\cdot \mathbf{F}_{\nu\mu}^{(\Phi)\alpha i,\beta j})\delta(\mathbf{r}_\mu^{\beta j}-\mathbf{r})\right\rangle$$

from 2nd $\langle\cdots\rangle$ term in Eq. (8.8)

$$=\tfrac{1}{2}\left\langle \sum_{\alpha i \nu}\sum_{\beta j \mu} ((\dot{\mathbf{r}}_\nu^{\alpha i}+\dot{\mathbf{r}}_\mu^{\beta j})\cdot \mathbf{F}_{\nu\mu}^{(\Phi)\alpha i,\beta j})\delta(\mathbf{r}_\nu^{\alpha i}-\mathbf{r})\right\rangle \tag{8.16}$$

We now consider the intramolecular contribution $Q^{(\phi)}$, that is, the terms for which $\alpha i = \beta j$. We apply Eqs. (5.9) and (5.10) and keep two terms, the first one of which is zero because of Newton's third law applied to two beads on one molecule:

$$Q^{(\phi)}=\tfrac{1}{2}\left\langle \sum_{\alpha i \nu\mu} ((\dot{\mathbf{r}}_\nu^{\alpha i}+\dot{\mathbf{r}}_\mu^{\alpha i})\cdot \mathbf{F}_{\nu\mu}^{(\phi)\alpha i})\delta(\mathbf{r}_\nu^{\alpha i}-\mathbf{r})\right\rangle$$

$$=\tfrac{1}{2}\sum_{\alpha\nu\mu}\int [[\dot{\mathbf{r}}_\nu^\alpha+\dot{\mathbf{r}}_\mu^\alpha]]^\alpha\cdot \mathbf{F}_{\nu\mu}^{(\phi)\alpha}\Psi_\alpha(\mathbf{r}-\mathbf{R}_\nu^\alpha,\mathbf{Q}^\alpha,t)d\mathbf{Q}^\alpha$$

$$=\tfrac{1}{2}\sum_{\alpha\nu\mu}\int [[\dot{\mathbf{r}}_\nu^\alpha+\dot{\mathbf{r}}_\mu^\alpha]]^\alpha\cdot \mathbf{F}_{\nu\mu}^{(\phi)\alpha}\Psi_\alpha(\mathbf{r},\mathbf{Q}^\alpha,t)d\mathbf{Q}^\alpha$$

$$-\nabla\cdot\tfrac{1}{2}\sum_{\alpha\nu\mu}\iint_0^1 \mathbf{R}_\nu^\alpha\mathbf{F}_{\nu\mu}^{(\phi)\alpha}\cdot[[\dot{\mathbf{r}}_\nu^\alpha+\dot{\mathbf{r}}_\mu^\alpha]]^\alpha\Psi_\alpha(\mathbf{r}-\xi\mathbf{R}_\nu^\alpha,\mathbf{Q}^\alpha,t)d\xi d\mathbf{Q}^\alpha$$

$$=-\nabla\cdot\tfrac{1}{2}\sum_{\alpha\nu\mu}\iint_0^1 \mathbf{R}_\nu^\alpha\mathbf{F}_{\nu\mu}^{(\phi)\alpha}\cdot[[(\dot{\mathbf{r}}_\nu^\alpha-\mathbf{v})+(\dot{\mathbf{r}}_\mu^\alpha-\mathbf{v})]]^\alpha$$

$$\cdot\Psi_\alpha(\mathbf{r}-\xi\mathbf{R}_\nu^\alpha,\mathbf{Q}^\alpha,t)d\xi d\mathbf{Q}^\alpha$$

$$-\nabla\cdot\left[\sum_{\alpha\nu\mu}\iint_0^1 \mathbf{R}_\nu^\alpha\mathbf{F}_{\nu\mu}^{(\phi)\alpha}\Psi_\alpha(\mathbf{r}-\xi\mathbf{R}_\nu^\alpha,\mathbf{Q}^\alpha,t)d\xi d\mathbf{Q}^\alpha\cdot\mathbf{v}(\mathbf{r},t)\right]$$

$$=-(\nabla\cdot\mathbf{q}^{(\phi)})-(\nabla\cdot[\boldsymbol{\pi}^{(\phi)}\cdot\mathbf{v}]) \tag{8.17}$$

which serves to define the intramolecular potential contribution $\mathbf{q}^{(\phi)}$ to the heat flux vector; the corresponding contribution to the stress tensor $\boldsymbol{\pi}^{(\phi)}$ was defined in Eq. (7.15). The first term after the third equals sign is zero, and the remaining term is rewritten by adding and subtracting $\mathbf{v}(\mathbf{r}, t)$.

The expression for the intramolecular force contribution to the heat-flux vector is then:

$$\mathbf{q}^{(\phi)}=\tfrac{1}{2}\sum_{\alpha\nu\mu}\iint_0^1 \mathbf{R}_\nu^\alpha\mathbf{F}_{\nu\mu}^{(\phi)\alpha}\cdot[[(\dot{\mathbf{r}}_\nu^\alpha-\mathbf{v})+(\dot{\mathbf{r}}_\mu^\alpha-\mathbf{v})]]^\alpha\Psi_\alpha(\mathbf{r}-\xi\mathbf{R}_\nu^\alpha,\mathbf{Q}^\alpha,t)d\xi d\mathbf{Q}^\alpha$$

$$=\tfrac{1}{2}\sum_{\alpha\nu\mu}\int \mathbf{R}_{\mu\nu}^\alpha\mathbf{F}_{\nu\mu}^{(\phi)\alpha}\cdot[[\dot{\mathbf{r}}_\nu^\alpha-\mathbf{v}]]^\alpha\Psi_\alpha(\mathbf{r},\mathbf{Q}^\alpha,t)d\mathbf{Q}^\alpha+\cdots \tag{8.18}$$

The first expression is given in Table 2, and the first term in the Taylor expansion appears in Table 1. In the latter we have used Newton's third law for pairs of beads on one molecule and the definition $\mathbf{R}_{\mu\nu}^\alpha=\mathbf{r}_\nu^\alpha-\mathbf{r}_\mu^\alpha$ (cf. Eq. (2.20))

The term $\mathbf{q}^{(\phi)}$, when dotted into $\mathbf{n}$ and multiplied by dS, describes the rate at which work is done on the fluid on the positive side of dS by the fluid on the negative side. The dot product of $\mathbf{F}_{\nu\mu}^{(\phi)\alpha}$ with the diffusive velocity $[[\dot{\mathbf{r}}_\nu^\alpha - \mathbf{v}]]^\alpha$ is the rate of doing work associated with the springs in one polymer molecule. The dot product of $\mathbf{n}$ with $\mathbf{R}_{\mu\nu}^\alpha$ indicates that springs that make a small angle with the plane of dS contribute less in the averaging process than springs that make a large angle.

8.4 The Intermolecular Contribution to the Heat-Flux Vector

From Eq. (8.16) we get the intermolecular contribution to the heat flux vector (that is, the terms for which $\alpha i \neq \beta j$), which can be transformed by using Eq. (5.20); the first term drops out by Newton's third law:

$$
\begin{aligned}
Q^{(d)} &= \tfrac{1}{2}\left\langle \sum_{\alpha i\nu}\sum_{\beta j\mu} ((\dot{\mathbf{r}}_\nu^{\alpha i} + \dot{\mathbf{r}}_\mu^{\beta j})\cdot\mathbf{F}_{\nu\mu}^{(d)\alpha i,\beta j})\,\delta(\mathbf{r}_\nu^{\alpha i} - \mathbf{r})\right\rangle \\
&= \tfrac{1}{2}\sum_{\alpha\beta\nu\mu}\iiint [[\dot{\mathbf{r}}_\nu^\alpha + \dot{\mathbf{r}}_\mu^\beta]]^{\alpha\beta}\cdot\mathbf{F}_{\nu\mu}^{(d)\alpha\beta}\,\tilde{\Psi}_{\alpha\beta}\,d\mathbf{R}_{\alpha\beta}\,d\mathbf{Q}^\alpha\,d\mathbf{Q}^\beta \\
&\quad - \nabla\cdot\tfrac{1}{2}\sum_{\alpha\beta\nu\mu}\iiint\left(\mathbf{R}_\nu^\alpha - \frac{m_m^\beta}{m_m^\alpha + m_m^\beta}\mathbf{R}_{\alpha\beta}\right)[[\dot{\mathbf{r}}_\nu^\alpha + \dot{\mathbf{r}}_\mu^\beta]]^{\alpha\beta} \\
&\quad \cdot\mathbf{F}_{\nu\mu}^{(d)\alpha\beta}\,\tilde{\Psi}_{\alpha\beta}\,d\mathbf{R}_{\alpha\beta}\,d\mathbf{Q}^\alpha\,d\mathbf{Q}^\beta \\
&= -\nabla\cdot\tfrac{1}{2}\sum_{\alpha\beta\nu\mu}\iiint\left(\mathbf{R}_\nu^\alpha - \frac{m_m^\beta}{m_m^\alpha + m_m^\beta}\mathbf{R}_{\alpha\beta}\right)[[(\dot{\mathbf{r}}_\nu^\alpha - \mathbf{v}) + (\dot{\mathbf{r}}_\mu^\beta - \mathbf{v})]]^{\alpha\beta} \\
&\quad \cdot\mathbf{F}_{\nu\mu}^{(d)\alpha\beta}\,\tilde{\Psi}_{\alpha\beta}\,d\mathbf{R}_{\alpha\beta}\,d\mathbf{Q}^\alpha\,d\mathbf{Q}^\beta \\
&\quad - \nabla\cdot\tfrac{1}{2}\sum_{\alpha\beta\nu\mu}\iiint\left(\mathbf{R}_\nu^\alpha - \frac{m_m^\beta}{m_m^\alpha + m_m^\beta}\mathbf{R}_{\alpha\beta}\right)\mathbf{F}_{\nu\mu}^{(d)\alpha\beta}\,\tilde{\Psi}_{\alpha\beta}\,d\mathbf{R}_{\alpha\beta}\,d\mathbf{Q}^\alpha\,d\mathbf{Q}^\beta \\
&\quad \cdot\mathbf{v}(\mathbf{r},t) \\
&= -(\nabla\cdot\mathbf{q}^{(d)}) - (\nabla\cdot[\boldsymbol{\pi}^{(d)}\cdot\mathbf{v}])
\end{aligned} \tag{8.19}
$$

Here we have made the identification of the second term in the last line with the help of Eq. (7.17). The expression for $\mathbf{q}^{(d)}$ from Eq. (8.19) can be modified by the same procedure used in going from Eq. (7.17) to (7.18), to give:

$$
\begin{aligned}
\mathbf{q}^{(d)} &= \tfrac{1}{2}\sum_{\alpha\beta\nu\mu}\iiint(\mathbf{R}_\nu^\alpha - \tfrac{1}{2}\mathbf{R}_{\alpha\beta})[[(\dot{\mathbf{r}}_\nu^\alpha - \mathbf{v}) + (\dot{\mathbf{r}}_\mu^\beta - \mathbf{v})]]^{\alpha\beta} \\
&\quad \cdot\mathbf{F}_{\nu\mu}^{(d)\alpha\beta}\,\tilde{\Psi}_{\alpha\beta}\,d\mathbf{R}_{\alpha\beta}\,d\mathbf{Q}^\alpha\,d\mathbf{Q}^\beta
\end{aligned} \tag{8.20}
$$

Finally, this last expression is written again with the indices $\alpha\nu$ and $\beta\mu$ interchanged, and then the procedure used in going from Eq. (7.18) to (7.19) is used.

This gives finally:

$$\mathbf{q}^{(d)} = \tfrac{1}{2} \sum_{\alpha\beta\nu\mu} \int \mathbf{R}_{\mu\nu}^{\beta\alpha} \mathbf{F}_{\nu\mu}^{(d)\alpha\beta} \cdot [[\dot{\mathbf{r}}_{\nu}^{\alpha} - \mathbf{v}]]^{\alpha\beta} \tilde{\Psi}_{\alpha\beta}(\mathbf{r}, \mathbf{R}_{\alpha\beta}, \mathbf{Q}^{\alpha}, \mathbf{Q}^{\beta}, t)\, d\mathbf{R}_{\alpha\beta} d\mathbf{Q}^{\alpha} d\mathbf{Q}^{\beta} \tag{8.21}$$

The term $\mathbf{q}^{(d)}$, when dotted into $\mathbf{n}$ and multiplied by dS, describes the rate at which work is done on the fluid on the positive side of dS by the fluid on the negative side, as a result of bead-bead interactions between beads on different molecules. Further comments on the structure of this term are similar to those given just after Eq. (8.18).

When the various contributions to the divergence of the heat-flux vector are combined, we finally get the term $-(\nabla \cdot \mathbf{q})$ in Eq. (8.1), with $\mathbf{q} = \mathbf{q}^{(k)} + \mathbf{q}^{(\phi)} + \mathbf{q}^{(d)} + \mathbf{q}^{(e)}$. In addition, the "work terms" in Eqs. (8.10), (8.13), (8.17), and (8.19) may be combined to give $-(\nabla \cdot [\boldsymbol{\pi} \cdot \mathbf{v}])$ in Eq. (8.1). Note also that the external forces give rise to the terms $(\mathbf{v} \cdot \mathbf{G})$ and J in the energy equation.

The various contributions to the heat-flux vector are summarized in Tables 1 and 2, along with the analogous contributions to the stress tensor.

9 The Hydrodynamic Angular Momentum Equation [11a]

The law of conservation of angular momentum as applied to a system in which there is no intrinsic angular momentum is:

$$\frac{\partial}{\partial t}[\mathbf{r} \times \rho\mathbf{v}] = -[\nabla \cdot \mathbf{v}[\mathbf{r} \times \rho\mathbf{v}]] - [\nabla \cdot \{\boldsymbol{\pi} \times \mathbf{r}\}] + [\mathbf{r} \times \mathbf{G}] + \mathbf{T} \tag{9.1}$$

in which $\mathbf{T}$ is the external torque per unit volume applied to the fluid, and $\mathbf{G}$ is the external force per unit volume (see Eq. (7.1)).

If we form the cross product of the position vector $\mathbf{r}$ with the hydrodynamic equation of motion in Eq. (7.1), the following equation is obtained:

$$\frac{\partial}{\partial t}[\mathbf{r} \times \rho\mathbf{v}] = -[\nabla \cdot \mathbf{v}[\mathbf{r} \times \rho\mathbf{v}]] - [\nabla \cdot \{\boldsymbol{\pi} \times \mathbf{r}\}] + [\mathbf{r} \times \mathbf{G}] - [\boldsymbol{\varepsilon} : \boldsymbol{\pi}] \tag{9.2}$$

in which $\boldsymbol{\varepsilon}$ is the isotropic third-order tensor, whose components ε_{ijk} are called the permutation symbols. (Eq. (9.2) is given on p. 831 of Ref. [12] for a symmetric stress tensor, but $\{\mathbf{r} \times \mathbf{p}\}$ there should be replaced by $\{\mathbf{r} \times \mathbf{p}\}^{\dagger}$.)

When Eq. (9.2) is subtracted from Eq. (9.1), we get

$$\mathbf{T} = -[\boldsymbol{\varepsilon} : \boldsymbol{\pi}] \tag{9.3}$$

which shows that the antisymmetrical part of the stress tensor is simply related to the external torque.

We now turn to the molecular derivation of the equation of change for angular momentum. The quantity $[\mathbf{r}_\nu^{\alpha i} \times \mathbf{p}_\nu^{\alpha i}]$ is the angular momentum of a bead with respect to some arbitrarily chosen fixed reference frame. The beads are regarded as point particles, and hence possess no intrinsic angular momentum. Consequently, to obtain Eq. (9.1) from the statistical mechanical approach, we consider the following vector function $\mathbf{B}'$ in the phase space:

$$\mathbf{B}' = \sum_{\alpha i \nu} [\mathbf{r}_\nu^{\alpha i} \times \mathbf{p}_\nu^{\alpha i}]\,\delta(\mathbf{r}_\nu^{\alpha i} - \mathbf{r}) \tag{9.4}$$

Then $\langle \mathbf{B}' \rangle$, which is the density of total angular momentum, is:

$$\begin{aligned}\langle \mathbf{B}' \rangle = \left\langle \sum_{\alpha i \nu} [\mathbf{r}_\nu^{\alpha i} \times \mathbf{p}_\nu^{\alpha i}]\,\delta(\mathbf{r}_\nu^{\alpha i} - \mathbf{r}) \right\rangle &= \mathbf{r} \times \left\langle \sum_{\alpha i \nu} \mathbf{p}_\nu^{\alpha i}\,\delta(\mathbf{r}_\nu^{\alpha i} - \mathbf{r}) \right\rangle \\ &= [\mathbf{r} \times \langle \mathbf{B} \rangle] = [\mathbf{r} \times \rho \mathbf{v}]\end{aligned} \tag{9.5}$$

in which $\mathbf{B}$ is the function defined in Eq. (7.2). Note that, according to the comments in the paragraph before Eq. (6.7), in the present treatment the mass of a molecule is distributed in space, in contrast to the usual treatment of polyatomic gases [13]. It is for this reason that no intrinsic angular momentum appears in Eq. (9.5).

Similarly, we find that

$$\langle \mathbb{L}\mathbf{B}' \rangle = [\mathbf{r} \times \langle \mathbb{L}\mathbf{B} \rangle] \tag{9.6}$$

Hence the general equation of change in Eq. (3.7) gives Eq. (9.2) directly.

In Eq. (9.2) the term $-[\boldsymbol{\varepsilon} : \boldsymbol{\pi}]$ would be identically zero if the stress tensor were symmetric. It is seen from Eq. (7.8) that the kinetic contribution to the stress tensor is symmetric; furthermore, as long as we assume that the bead-bead interaction forces are collinear with the bead-bead vectors, then the intramolecular and intermolecular contributions to the stress tensor in Eqs. (7.15) and (7.19) must also be symmetric. The external force contribution in Eq. (7.12) is in general not symmetric. For further discussions regarding this point, see Appendix A.

It can be concluded, then, that when the external forces per unit mass are different for the various beads of a particular species, the term $-[\boldsymbol{\varepsilon} : \boldsymbol{\pi}]$ in Eq. 9.2 will in general not vanish. Note also that the quantity $\{\boldsymbol{\pi} \times \mathbf{r}\} = \{\mathbf{r} \times \boldsymbol{\pi}^\dagger\}^\dagger$ represents the angular momentum flux tensor.

10 Equations for the Time-Evolution of the Singlet Distribution Functions (DPL, Sect. 17.5)

In order to use the flux expressions developed in the foregoing sections, it is necessary to have the singlet and doublet distribution functions. The partial differential equation for the singlet distribution function is discussed here. The analogous discussion for the doublet distribution function has been given by Curtiss [14].

We start by applying the general equation of change, Eq. (3.7), to the quantity:

$$B = \Sigma_i B_i = \Sigma_i \delta(\mathbf{r}^{\alpha i} - \mathbf{r}^{\alpha})\delta(\mathbf{p}^{\alpha i} - \mathbf{p}^{\alpha}) \tag{10.1}$$

Then, according to Eq. (4.1), $\langle B \rangle = f_\alpha(\mathbf{r}^\alpha, \mathbf{p}^\alpha, t)$ and

$$\frac{\partial}{\partial t} f_\alpha = \left\langle \sum_{\beta j \nu} \left(\frac{\mathbf{p}_\nu^{\beta j}}{m_\nu^\beta} \cdot \frac{\partial}{\partial \mathbf{r}_\nu^{\beta j}} \Sigma_i B_i + \mathbf{F}_\nu^{\beta j} \cdot \frac{\partial}{\partial \mathbf{p}_\nu^{\beta j}} \Sigma_i B_i \right) \right\rangle \tag{10.2}$$

Only those terms with $\beta = \alpha$ and $j = i$ contribute, so that:

$$\frac{\partial}{\partial t} f_\alpha = - \sum_{i\nu} \left(\frac{1}{m_\nu^\alpha} \frac{\partial}{\partial \mathbf{r}_\nu^\alpha} \cdot \langle \mathbf{p}_\nu^{\alpha i} B_i \rangle + \frac{\partial}{\partial \mathbf{p}_\nu^\alpha} \cdot \langle \mathbf{F}_\nu^{\alpha i} B_i \rangle \right) \tag{10.3}$$

In $\langle \mathbf{p}_\nu^{\alpha i} B_i \rangle$, because of the integrals over the delta functions, $\mathbf{p}_\nu^{\alpha i}$ may be replaced by $\mathbf{p}_\nu^\alpha$. The same thing can be done for the terms involving $\langle \mathbf{F}_\nu^{(e)\alpha i} B_i \rangle$ and $\langle \mathbf{F}_\nu^{(\phi)\alpha i} B_i \rangle$ in the last term. The contribution $\langle \mathbf{F}_\nu^{(d)\alpha i} B_i \rangle = \langle \Sigma_\beta \Sigma_j \Sigma_\mu \mathbf{F}_{\nu\mu}^{(d)\alpha i, \beta j} B_i \rangle$ requires special treatment. First, integrals over the delta functions $\delta(\mathbf{r}^{\beta j} - \mathbf{r}^\beta)$ and $\delta(\mathbf{p}^{\beta j} - \mathbf{p}^\beta)$ are inserted inside the angular brackets (cf. Sect. 5) and then use is made of Eq. (4.6):

$$\begin{aligned} &\sum_i \sum_{\beta j \mu} \iiint \mathbf{F}_{\nu\mu}^{(d)\alpha i, \beta j} \delta(\mathbf{r}^{\alpha i} - \mathbf{r}^\alpha)\, \delta(\mathbf{p}^{\alpha i} - \mathbf{p}^\alpha) \\ &\qquad \cdot \delta(\mathbf{r}^{\beta j} - \mathbf{r}^\beta) \delta(\mathbf{p}^{\beta j} - \mathbf{p}^\beta) f(x, t) d\mathbf{r}^\beta\, d\mathbf{p}^\beta\, dx \\ &= \sum_{\beta\mu} \iint \mathbf{F}_{\nu\mu}^{(d)\alpha\beta} \langle \Sigma_i \Sigma_j \delta(\mathbf{r}^{\alpha i} - \mathbf{r}^\alpha) \delta(\mathbf{p}^{\alpha i} - \mathbf{p}^\alpha) \\ &\qquad \cdot \delta(\mathbf{r}_\nu^{\beta j} - \mathbf{r}^\beta) \delta(\mathbf{p}^{\beta j} - \mathbf{p}^\beta) \rangle d\mathbf{r}^\beta\, d\mathbf{p}^\beta \\ &= \sum_{\beta\mu} \iint \mathbf{F}_{\nu\mu}^{(d)\alpha\beta} f_{\alpha\beta}(\mathbf{r}^\alpha, \mathbf{p}^\alpha, \mathbf{r}^\beta, \mathbf{p}^\beta, t) d\mathbf{r}^\beta\, d\mathbf{p}^\beta \end{aligned} \tag{10.4}$$

When this is inserted into the previous equation, we finally get the equation for the time evolution of the distribution function in the one-molecule phase space:

$$\begin{aligned} \frac{\partial}{\partial t} f_\alpha = &- \Sigma_\nu \left(\frac{\mathbf{p}_\nu^\alpha}{m_\nu^\alpha} \cdot \frac{\partial}{\partial \mathbf{r}_\nu^\alpha} f_\alpha + \mathbf{F}_\nu^{(e)\alpha} \cdot \frac{\partial}{\partial \mathbf{p}_\nu^\alpha} f_\alpha + \mathbf{F}_\nu^{(\phi)\alpha} \cdot \frac{\partial}{\partial \mathbf{p}_\nu^\alpha} f_\alpha \right) \\ &- \Sigma_\nu \left(\frac{\partial}{\partial \mathbf{p}_\nu^\alpha} \cdot \sum_{\beta\mu} \iint \mathbf{F}_{\nu\mu}^{(d)\alpha\beta} f_{\alpha\beta}(\mathbf{r}^\alpha, \mathbf{p}^\alpha, \mathbf{r}^\beta, \mathbf{p}^\beta, t) d\mathbf{r}^\beta\, d\mathbf{p}^\beta \right) \end{aligned} \tag{10.5}$$

Integration over all momenta $\mathbf{p}_\nu^\alpha$, and use of the definitions in Eq. (4.3) and (5.1), then give for the time evolution of $\bar{\Psi}_\alpha(\mathbf{r}^\alpha, t)$:

$$\frac{\partial}{\partial t} \bar{\Psi}_\alpha(\mathbf{r}^\alpha, t) = - \Sigma_\nu \left(\frac{\partial}{\partial \mathbf{r}_\nu^\alpha} \cdot [[\dot{\mathbf{r}}_\nu^\alpha]]^\alpha \bar{\Psi}_\alpha \right) \tag{10.6}$$

and, when Eq. (2.13) is used, we get the equation for $\Psi_\alpha(\mathbf{r}_c^\alpha, \mathbf{Q}^\alpha, t)$:

$$\frac{\partial}{\partial t}\Psi_\alpha(\mathbf{r}_c^\alpha, \mathbf{Q}^\alpha, t) = -\left(\frac{\partial}{\partial \mathbf{r}_c^\alpha}\cdot[[\dot{\mathbf{r}}_c^\alpha]]^\alpha \Psi_\alpha\right) - \Sigma_j\left(\frac{\partial}{\partial \mathbf{Q}_j^\alpha}\cdot[[\dot{\mathbf{Q}}_j^\alpha]]^\alpha \Psi_\alpha\right) \tag{10.7}$$

The $[[\ ]]^\alpha$ quantities in these last two equations have the same arguments as the distribution functions with which they are associated. Note that Eqs. (10.5), (10.6), and (10.7) are valid for any species in a multicomponent mixture involving flexible macromolecules. Equation (10.7) is a generalization of the usual "equation of continuity" for $\Psi_\alpha(\mathbf{r}_c^\alpha, \mathbf{Q}^\alpha, t)$ for Rouse chains (DPL, Eqs. (15.1–5)), in that it is applicable to models of any connectivity and with bead masses and friction coefficients different from one another. One often sees the equation written without the momentum-space averages for the velocities; in such instances the equation contains an inappropriate mixture of statistical and deterministic quantities.

Eq. (10.6) – or (10.7) – cannot be used to get the distribution function until more is known about the momentum-averaged quantities. In Sect. 12.4 an empirical expression is presented for these quantities, and some examples of solving the resulting "diffusion equations" are shown in Sect. 13.

11 Equations of Internal Motion for the Molecules; Hydrodynamic and Brownian Forces (DPL, Sect. 17.5)

To get the (statistically averaged) equations of motion for the beads, we multiply Eq. (10.5) by $\mathbf{p}_\nu^{\alpha i}$ and integrate over all the momenta of molecule α. This gives, when use is made of Eqs. (4.3), (4.7), and (5.1):

$$m_\nu^\alpha \frac{\partial}{\partial t}[[\dot{\mathbf{r}}_\nu^\alpha]]^\alpha \bar{\Psi}_\alpha = -m_\nu^\alpha \Sigma_\mu\left(\frac{\partial}{\partial \mathbf{r}_\mu^\alpha}\cdot[[\dot{\mathbf{r}}_\mu^\alpha \dot{\mathbf{r}}_\nu^\alpha]]^\alpha \bar{\Psi}_\alpha\right)$$
$$+ \mathbf{F}_\nu^{(e)\alpha}\bar{\Psi}_\alpha + \mathbf{F}_\nu^{(\phi)\alpha}\bar{\Psi}_\alpha + \Sigma_\beta \int \mathbf{F}_\nu^{(d)\alpha\beta}\bar{\Psi}_{\alpha\beta}\,d\mathbf{r}^\beta \tag{11.1}$$

Next we replace the double bracket in the first term on the right side by $[[(\dot{\mathbf{r}}_\mu^\alpha - \mathbf{u}_\mu^\alpha)(\dot{\mathbf{r}}_\nu^\alpha - \mathbf{u}_\nu^\alpha)]]^\alpha$, and add appropriate compensating terms; here and elsewhere we use the notation $\mathbf{u}_\nu^{\alpha i}(\mathbf{r}^\alpha, t) = [[\dot{\mathbf{r}}_\nu^\alpha]]^\alpha$ for the average velocity of bead ν. Then Eq. (11.1) becomes:

$$m_\nu^\alpha \frac{\partial}{\partial t}\mathbf{u}_\nu^\alpha \bar{\Psi}_\alpha + m_\nu^\alpha \Sigma_\mu\left(\frac{\partial}{\partial \mathbf{r}_\mu^\alpha}\cdot \mathbf{u}_\mu^\alpha \mathbf{u}_\nu^\alpha \bar{\Psi}_\alpha\right)$$
$$= -m_\nu^\alpha \Sigma_\mu\left(\frac{\partial}{\partial \mathbf{r}_\mu^\alpha}\cdot[[(\dot{\mathbf{r}}_\mu^\alpha - \mathbf{u}_\mu^\alpha)(\dot{\mathbf{r}}_\nu^\alpha - \mathbf{u}_\nu^\alpha)]]^\alpha \bar{\Psi}_\alpha\right)$$
$$+ \mathbf{F}_\nu^{(e)\alpha}\bar{\Psi}_\alpha + \mathbf{F}_\nu^{(\phi)\alpha}\bar{\Psi}_\alpha + \Sigma_\beta \int \mathbf{F}_\nu^{(d)\alpha\beta}\bar{\Psi}_{\alpha\beta}\,d\mathbf{r}^\beta \tag{11.2}$$

The terms on the left side can be differentiated by parts, and Eq. (10.6) can then be used to show that two of the terms sum to zero. Thus, the left side of Eq. (11.2) becomes:

$$m_\nu^\alpha \bar{\Psi}_\alpha \frac{\partial}{\partial t} \mathbf{u}_\nu^\alpha + m_\nu^\alpha \bar{\Psi}_\alpha \Sigma_\mu \left(\mathbf{u}_\mu^\alpha \cdot \frac{\partial}{\partial \mathbf{r}_\mu^\alpha} \mathbf{u}_\nu^\alpha \right) \tag{11.3}$$

When Eq. (11.2) is then divided by $\bar{\Psi}_\alpha$ we get the equation of motion for the beads:

$$m_\nu^\alpha \left(\frac{\partial}{\partial t} \mathbf{u}_\nu^\alpha + \Sigma_\mu \left(\mathbf{u}_\mu^\alpha \cdot \frac{\partial}{\partial \mathbf{r}_\mu^\alpha} \mathbf{u}_\nu^\alpha \right) \right) = \mathbf{F}_\nu^{(b)\alpha} + \mathbf{F}_\nu^{(e)\alpha} + \mathbf{F}_\nu^{(\phi)\alpha} + \mathbf{F}_\nu^{(h)\alpha} \tag{11.4}$$

We refer to the terms on the left as "acceleration terms," and on the right we have a sum of four forces. There are two forces $\mathbf{F}_\nu^{(b)\alpha}$ and $\mathbf{F}_\nu^{(h)\alpha}$ whose sum is well determined by comparing Eqs. (11.2) and (11.4). However, the separation into two terms is arbitrary. The symbols $\mathbf{F}_\nu^{(b)\alpha}$ and $\mathbf{F}_\nu^{(h)\alpha}$ respectively are referred to here as the *averaged Brownian force* and the *averaged hydrodynamic force* on bead ν of molecule α. We choose to define these forces, in the most natural way, as follows:

$$\mathbf{F}_\nu^{(b)\alpha}(\mathbf{r}^\alpha, t) = -\frac{1}{\bar{\Psi}_\alpha} \Sigma_\mu \left(\frac{\partial}{\partial \mathbf{r}_\mu^\alpha} \cdot m_\nu^\alpha [[(\dot{\mathbf{r}}_\mu^\alpha - \mathbf{u}_\mu^\alpha)(\dot{\mathbf{r}}_\nu^\alpha - \mathbf{u}_\nu^\alpha)]]^\alpha \bar{\Psi}_\alpha \right) \tag{11.5}$$

$$\mathbf{F}_\nu^{(h)\alpha}(\mathbf{r}^\alpha, t) = \frac{1}{\bar{\Psi}_\alpha} \Sigma_{\beta\mu} \int \mathbf{F}_{\nu\mu}^{(d)\alpha\beta} \bar{\Psi}_{\alpha\beta} d\mathbf{r}^\beta \tag{11.6}$$

The identification of the "averaged Brownian force" in Eq. (11.5) seems plausible, since the integrations over momenta using a Maxwellian distribution about the velocity $\mathbf{u}_\nu^\alpha$ lead to the standard formula for the Brownian force that one finds in most kinetic theory discussions (See Sect. 12.5). Furthermore it is not altogether unreasonable to refer to the force in Eq. (11.6) as the "averaged hydrodynamic force", since it involves the sum of all the intermolecular forces acting on the bead. To date the right side of Eq. (11.6) has not been evaluated for any molecular models or flow situations; it has been standard practice to represent the hydrodynamic force by an empirical expression involving a "friction coefficient," and we will follow that custom here (See Sect. 12.4). (Note: In DPL the identifications of acceleration terms and the Brownian force are somewhat different from those made here.)

It should be noted that Eq. (11.4) has some similarity to the Langevin equation for a bead of a polymer molecule. However, the Langevin equation is effectively a statement of Newton's second law of motion, containing a stochastic term to represent the randomness of the Brownian force. Equation (11.4), on the other hand, is a statistical equation of motion involving averaged quantities.

12 Five Assumptions

Up to this point the treatment has been fairly general. The only assumptions involved are:

a. The molecular model of beads and springs (that is, a constraint-free model).
b. The assumption of pairwise additive potentials between all beads in the system.

An additional approximation that may be made is the systematic omission of higher-order terms in the flux expressions, because of truncations of the Taylor series mentioned in connection with Eqs. (5.10) and (5.20). This implies the assumption that n_α, $\mathbf{v}$, and T, and their spatial derivatives, do not change significantly over molecular dimensions. The results in Table 1 are therefore frequently appropriate starting points for the study of the rheology, diffusion, and heat conduction in flexible polymer mixtures, both in solutions and melts. However, for the study of "cross effects" the first three terms in the Taylor series are needed, as discussed in Sects 14–16.

In this section we discuss five assumptions that have traditionally been introduced in order to continue with the development of the kinetic theory of transport phenomena and get useful results. Some of these assumptions are made because we do not at present know enough about the distribution functions that appear in the expressions in Tables 1 and 2, in particular the pair distribution function and the momentum-space distribution function. Other assumptions are introduced in order to simplify the subsequent problem-solving for specific molecular models. All the assumptions we present here can be challenged; some of them should be modified, and some of them may ultimately be eliminated.

12.1 *The Short-Range-Force Assumption*

In Table 1 it can be seen that the intermolecular force contributions to the stress tensor and heat flux involve the forces between beads on different molecules, $\mathbf{F}_{\nu\mu}^{(d)\alpha\beta}$. We assume that this force is negligibly small unless bead ν of molecule α and bead μ of molecule β are extremely close (see Fig. (18.1-1) of DPL). That is, we state that

$$\mathbf{F}_{\nu\mu}^{(d)\alpha\beta} = \mathbf{0} \text{ unless } \mathbf{R}_{\nu\mu}^{\alpha\beta} = \mathbf{0} \tag{12.1}$$

which is referred to as the *short-range-force assumption.* This seems to be similar to the notion in transient network theories (See DPL, Chap. 20) that there exist "temporary junctions" where segments of two proximate polymer chains interact strongly.

When this assumption is made, the intermolecular force contributions in Eqs. (7.19) and (8.21) are both zero. This means that there is no need to know the pair distribution function that appears in these two expressions. It should be noted, however, that the pair distribution function also appears in Eq. (8.11) for the kinetic contribution to the heat flux vector. One can argue that, if the short-range force approximation is made, then the third term in Eq. (8.11) can be neglected: the potentials $\phi_{\nu\mu}^{(d)\alpha\beta}$ are regarded as being zero, except for extremely small distances (of the order of the "collision diameter" of the beads), and the pair distribution function is zero when the beads overlap.

The above reasoning has been used to eliminate the need for knowing the pair distribution function in the kinetic theory of polymer melts [9, 14a].

For dilute solutions, one generally considers the solvent to be a continuum, and polymer-solvent interactions are not explicitly considered. As a result the terms containing the pair distribution functions (in Eqs. (7.19), (8.11), and (8.21)) are not needed, since polymer-polymer interactions occur only rarely. On the other hand, if one wishes to study polymer-solvent effects explicitly, then it will be necessary to consider in detail the terms containing the pair distribution function.

12.2 The Assumption of Linear Gradients

In Sects. 6, 7, and 8 we have derived equations for the conservation of linear momentum, energy and the mass of molecular species. We now define the gradients of the related variables as: $\boldsymbol{\kappa} = (\nabla \mathbf{v})^\dagger$, and $\mathbf{a} = \nabla \ln T$, and $\mathbf{b}_\alpha = \nabla \ln n_\alpha$. In general each of these gradients is a function of both position $\mathbf{r}_c$ and time t. It is usually adequate to assume that the higher derivatives of these variables are sufficiently small that, over distances comparable to molecular dimensions, they may be neglected, and thus we may use the following truncated Taylor series for velocity, temperature, and concentration:

$$\mathbf{v}(\mathbf{r}_\nu^\alpha, t) = \mathbf{v}(\mathbf{r}_c^\alpha, t) + [(\mathbf{r}_\nu^\alpha - \mathbf{r}_c^\alpha)\cdot\nabla\mathbf{v}] = \mathbf{v}(\mathbf{r}_c^\alpha, t) + [\boldsymbol{\kappa}\cdot\mathbf{R}_\nu^\alpha] \tag{12.2}$$

$$T(\mathbf{r}_\nu^\alpha, t) = T(\mathbf{r}_c^\alpha, t) + ((\mathbf{r}_\nu^\alpha - \mathbf{r}_c^\alpha)\cdot\nabla T) = T(\mathbf{r}_c^\alpha, t)[1 + (\mathbf{a}\cdot\mathbf{R}_\nu^\alpha)] \tag{12.3}$$

$$n_\alpha(\mathbf{r}_\nu^\alpha, t) = n_\alpha(\mathbf{r}_c^\alpha, t) + ((\mathbf{r}_\nu^\alpha - \mathbf{r}_c^\alpha)\cdot\nabla n_\alpha) = n_\alpha(\mathbf{r}_c^\alpha, t)[1 + (\mathbf{b}_\alpha\cdot\mathbf{R}_\nu^\alpha)] \tag{12.4}$$

With this assumption each molecule sees only constant gradients of velocity, temperature, and concentration.

The assumption that the gradients are constant over molecular dimensions (except near walls) appears to be reasonable. However, it is pointed out later, in Sect. 13, that the systems being considered here have "memory" and that the singlet distribution functions for a nonequilibrium system at a point depend on the "past history" of the molecules as they move through space. In particular, the solution of the equation for the distribution function requires a knowledge of the gradients along fluid particle path lines over times sufficiently long that the distances may be significantly larger than molecular dimensions. Thus it is usually necessary to assume that the gradients are constant in space (but not time) over these longer distances. This is the assumption of *homogeneous flow*.

In considering isothermal systems of constant composition, we set both $\mathbf{a}$ and $\mathbf{b}_\alpha$ equal to zero. In later discussions, however, we will sometimes assume that the gradients $\mathbf{a}$ and $\mathbf{b}_\alpha$ are small, and then retain only those terms linear in the temperature and concentration gradients.

12.3 The Assumption of Negligible Acceleration Terms

In the equations of motion for the beads in Eq. (11.4), it seems reasonable to assume that the bead-acceleration terms on the left side (which have the general appearance of a "substantial derivative") are negligibly small compared to the individual terms on the right side. This assumption has been discussed in several publications [15–19]. When the acceleration terms are omitted, the bead equation of motion reduces to a simple force balance used in the publications of Kramers [1], Kirkwood [2], Rouse [3], Zimm [4], and others:

$$\mathbf{F}_\nu^{(b)\alpha} + \mathbf{F}_\nu^{(e)\alpha} + \mathbf{F}_\nu^{(\phi)\alpha} + \mathbf{F}_\nu^{(h)\alpha} = 0 \tag{12.5}$$

That is, the sum of the Brownian, external, intramolecular, and hydrodynamic forces acting on a bead is zero. The introduction of this assumption makes the solution of the equation for the singlet configuration-space distribution function simpler. It is also useful in developing alternative expressions for the stress tensor (see, e.g., DPL Tables 13.3-1 and 15.2-1).

12.4 An Assumption for the Hydrodynamic Force – Introduction of the Friction Coefficient

In Eq. (11.6) an expression is given for the hydrodynamic force on a bead. This expression involves the pair distribution function, which is not known. Therefore it has been standard practice to make use of an empiricism. In its simplest form, this states that the average hydrodynamic force on a bead depends linearly on the difference between the average bead velocity and the velocity of the surrounding medium (Kuhn [20], Kramers [1], Kirkwood and coworkers [2]):

$$\mathbf{F}_\nu^{(h)\alpha}(\mathbf{r}^\alpha, t) = -\zeta_\nu^\alpha([[\dot{\mathbf{r}}_\nu^\alpha]]^\alpha - \mathbf{v}(\mathbf{r}_\nu^\alpha, t)) = -\zeta_\nu^\alpha(\mathbf{u}_\nu^\alpha(\mathbf{r}^\alpha, t) - \mathbf{v}(\mathbf{r}_\nu^\alpha, t)) \tag{12.6}$$

in which ζ_ν^α is the *friction coefficient* for bead ν of molecule α. When this empiricism is combined with the assumption that the velocity gradient is constant over molecular dimensions, and when the force balance is used, we obtain for the momentum-space-averaged bead velocity:

$$[[\dot{\mathbf{r}}_\nu^\alpha]]^\alpha \equiv \mathbf{u}_\nu^\alpha(\mathbf{r}^\alpha, t) = \mathbf{v}(\mathbf{r}_c^\alpha, t) + [\boldsymbol{\kappa}\cdot\mathbf{R}_\nu^\alpha] - \frac{1}{\zeta_\nu^\alpha}\mathbf{F}_\nu^{(h)\alpha}(\mathbf{r}^\alpha, t)$$

$$= \mathbf{v}(\mathbf{r}_c^\alpha, t) + [\boldsymbol{\kappa}\cdot\mathbf{R}_\nu^\alpha] + \frac{1}{\zeta_\nu^\alpha}(\mathbf{F}_\nu^{(b)\alpha} + \mathbf{F}_\nu^{(e)\alpha} + \mathbf{F}_\nu^{(\phi)\alpha}) \tag{12.7}$$

This expression is used in the next section in developing the equation for the singlet configuration-space distribution function.

Just before Eq. (12.6) it was stated that this equation is the simplest empiricism that has been suggested for the hydrodynamic force. Other empiricisms that have been used are:

i. The inclusion of *hydrodynamic interaction*: According to Eq. (12.6), the fluid velocity in the neighborhood of the bead is that of the imposed flow field. A better description allows for the perturbation of the flow field in dilute solutions, resulting from the motions of all the other beads of the molecule [2, 4]. There are at least two empiricisms available for inclusion of this effect: the Oseen-Burgers expression (DPL, pp. 94, 164), and the Rotne-Prager-Yamakawa expression (DPL, p. 94). The need for including hydrodynamic interaction in the theory for dilute solutions is evident from the very good description of the molecular weight dependence of linear viscoelastic properties and the translational diffusivity (DPL, Sect. 15.4). For a more extensive discussion see DPL, Sect. 18.2.

ii. The inclusion of a *tensorial friction coefficient*: It is not necessary to require that the hydrodynamic force be in the same direction as the relative bead velocity as in Eq. (12.6). It has been suggested that, in concentrated solutions and undiluted polymers (modeled by bead–spring chains), the beads can move more easily in the direction along the backbone than in the direction perpendicular to it. This idea can be expressed by allowing the friction coefficient to be a second-order tensor (DPL Sects. 13.1, 13.7, and 19.1); this introduces additional undetermined parameters into the theory [9, 20a, 20b, 20c].

iii. The inclusion of *pairwise frictional forces*: Eq. (12.6) involves the difference between the averaged bead velocity and the fluid velocity. As seen in Appendix B this empiricism leads to an inconsistency, in that the mass fluxes $\mathbf{j}_\alpha$ in Eq. (B.17) do not sum to zero as required by the definition in Eq. (6.4). This inconsistency has led to replacing Eq. (12.6) by a different empiricism that accounts for the interactions between all pairs of species (see Ref. [20d]).

The idea of introducing a friction coefficient into kinetic theory has a long history. The early papers by Kirkwood and collaborators [2], as well as the more recent work by Curtiss [14], should be consulted regarding the justification for the use of this empiricism. In most kinetic theories the friction coefficient ends up as an undetermined parameter, appearing – along with the spring constant – in the expression for a time constant (e.g., as $\lambda_H = \zeta/4H$ in the Hookean dumbbell model, and as $\lambda_j = \zeta c_j/4H$ in the Rouse model (see Eqs. (13.6) and (13.9)). The time constants are then determined from experimental data on rheological properties, such as viscosity, complex viscosity, or normal stresses.

12.5 *An Assumption for the Brownian Force – Introduction of the Temperature*

We define a local temperature $T_\nu^\alpha(\mathbf{r}, t)$, associated with beads ν of species α at point $\mathbf{r}$ at time t, in terms of the average kinetic energy of these beads with respect to $\mathbf{u}_\nu^\alpha$ as:

$$\tfrac{3}{2}\, kT_\nu^\alpha(\mathbf{r}, t) = \frac{\langle \Sigma_i \tfrac{1}{2} m_\nu^\alpha (\dot{\mathbf{r}}_\nu^{\alpha i} - \mathbf{u}_\nu^\alpha)^2\, \delta(\mathbf{r}_\nu^{\alpha i} - \mathbf{r})\rangle}{\langle \Sigma_i \delta(\mathbf{r}_\nu^{\alpha i} - \mathbf{r})\rangle} \tag{12.8}$$

Here, k is the Boltzmann constant, and $\mathbf{u}_\nu^\alpha$ is the same function as defined in Eq. (12.7), but with all the $\mathbf{r}_\mu^\alpha$ replaced by $\mathbf{r}_\mu^{\alpha i}$. From the definition, Eq. (12.8), and Eq. (5.9), it follows that

$$\tfrac{3}{2}\, kT_\nu^\alpha(\mathbf{r}, t) = \frac{\tfrac{1}{2} m_\nu^\alpha \int [[(\dot{\mathbf{r}}_\nu^\alpha - \mathbf{u}_\nu^\alpha)^2]]^\alpha \Psi_\alpha(\mathbf{r} - \mathbf{R}_\nu^\alpha, \mathbf{Q}^\alpha, t)\, d\mathbf{Q}^\alpha}{\int \Psi_\alpha(\mathbf{r} - \mathbf{R}_\nu^\alpha, \mathbf{Q}^\alpha, t)\, d\mathbf{Q}^\alpha} \tag{12.9}$$

where now $\mathbf{u}_\nu^\alpha$ is a function of the arguments of Ψ_α.

In Eq. (11.5) an expression for the Brownian force acting on a bead was given. In order to use this expression, some proposal has to be made for evaluating the momentum-space averages. To do this we begin by defining a function $\Xi_\alpha(\mathbf{r}^\alpha, \mathbf{p}^\alpha, t)$ by:

$$f_\alpha(\mathbf{r}^\alpha, \mathbf{p}^\alpha, t) = \bar{\Psi}_\alpha(\mathbf{r}^\alpha, t)\, \Xi_\alpha(\mathbf{r}^\alpha, \mathbf{p}^\alpha, t) \tag{12.10}$$

so that $\Xi_\alpha(\mathbf{r}^\alpha, \mathbf{p}^\alpha, t)$ describes the distribution in the momentum space for fixed values of the $\mathbf{r}_\nu^\alpha$.

We take the function $\Xi_\alpha(\mathbf{r}^\alpha, \mathbf{p}^\alpha, t)$ to be a product of Maxwellian distributions, one for each bead about its own average momentum:

$$\Xi_\alpha(\mathbf{r}^\alpha, \mathbf{p}^\alpha, t) = \frac{\exp[-\Sigma_\lambda \beta_\lambda^\alpha (\mathbf{p}_\lambda^\alpha - [[\mathbf{p}_\lambda^\alpha]]^\alpha)^2]}{\int \exp[-\Sigma_\lambda \beta_\lambda^\alpha (\mathbf{p}_\lambda^\alpha - [[\mathbf{p}_\lambda^\alpha]]^\alpha)^2]\, d\mathbf{p}^\alpha} \tag{12.11}$$

in which the mean momentum is $[[\mathbf{p}_\lambda^\alpha]]^\alpha = m_\lambda^\alpha \mathbf{u}_\lambda^\alpha(\mathbf{r}^\alpha, t)$. This distribution functions is normalized to unity, so as to be consistent with the normalization condition for $f_\alpha(\mathbf{r}^\alpha, \mathbf{p}^\alpha, t)$ in Eq. (4.2) and with the definition of $\bar{\Psi}_\alpha(\mathbf{r}^\alpha, t)$ in Eq. (4.3).

The functions β_λ^α in the configuration space are determined by the local temperatures through the definition in Eq. (12.8) or Eq. (12.9). It follows from the definitions in Eq. (5.1) and Eq. (12.10) that:

$$[[(\dot{\mathbf{r}}_\nu^\alpha - \mathbf{u}_\nu^\alpha)^2]]^\alpha = \frac{3}{2(\mathbf{m}_\nu^\alpha)^2 \beta_\nu^\alpha} \tag{12.12}$$

When this is substituted into Eq. (12.9), we find that to be consistent with the definition of the local temperatures, the β_λ^α must be chosen so that:

$$\tfrac{3}{2}\, kT_\nu^\alpha(\mathbf{r}, t) = \frac{3}{4m_\nu^\alpha} \frac{\int (1/\beta_\nu^\alpha) \Psi_\alpha(\mathbf{r} - \mathbf{R}_\nu^\alpha, \mathbf{Q}^\alpha, t)\, d\mathbf{Q}^\alpha}{\int \Psi_\alpha(\mathbf{r} - \mathbf{R}_\nu^\alpha, \mathbf{Q}^\alpha, t)\, d\mathbf{Q}^\alpha} \tag{12.13}$$

in which β_ν^α is a function of the arguments of the distribution function. We now choose the β_λ^α to be of the form:

$$\beta_\lambda^\alpha = \frac{1}{2m_\lambda^\alpha k T_\lambda^\alpha(\mathbf{r}_\lambda^\alpha, t)} \tag{12.14}$$

where T_λ^α is the absolute temperature at a bead λ at time t. When this is substituted into Eq. (12.13) and evaluated at the displaced coordinate, so that:

$$\beta^{\alpha}_{\lambda}|_{\mathbf{r}-\mathbf{R}^{\alpha}_{\nu},\mathbf{Q}^{\alpha},t} = \frac{1}{2m^{\alpha}_{\lambda}kT^{\alpha}_{\lambda}(\mathbf{r},t)} \tag{12.15}$$

the condition of Eq. (12.13) is satisfied.

It is now further assumed that the functions T^{α}_{λ} for all beads of all species are identical functions of position of time, namely $T(\mathbf{r}, t)$. This assumption, along with Eq. (12.11), will be referred to as the *assumption of equilibration in momentum space*. That is, we assume that the distribution function in momentum space for a particular bead is the same as that for a bead in a system at equilibrium at the fluid temperature surrounding the bead in question. By allowing for the variation of the temperature over the full extent of a molecule, we are then in a position to study non-isothermal problems--that is, situations in which $\mathbf{a} = \nabla \ln T$ is nonzero.

We point out parenthetically that in the kinetic theory of dilute gases it is just the deviation from the Maxwellian velocity distribution that is of primary interest in the evaluation of the transport properties. In the kinetic theory of polymers, on the other hand, it has been assumed that the deviations from the Maxwellian distribution are of minor importance, and to date few calculations or estimations have been made of the errors introduced by this assumption [16–19]. These exploratory efforts indicate that there may be a significant effect on the components of the complex viscosity in high-frequency oscillatory shearing flows.

When the distribution in Eq. (12.11) is substituted into Eq. (11.5) we get:

$$\begin{aligned}\mathbf{F}^{(b)\alpha}_{\nu}(\mathbf{r}^{\alpha}, t) &= -\frac{1}{\bar{\Psi}_{\alpha}}\frac{\partial}{\partial \mathbf{r}^{\alpha}_{\nu}}(kT(\mathbf{r}^{\alpha}_{\nu}, t)\bar{\Psi}_{\alpha}) \quad \text{Nonisothermal system} \\ &= -kT(\mathbf{r}_c)[1 + (\mathbf{a}\cdot\mathbf{R}^{\alpha}_{\nu})]\frac{\partial}{\partial \mathbf{r}^{\alpha}_{\nu}}\ln \bar{\Psi}_{\alpha} - kT(\mathbf{r}_c)\,\mathbf{a}\end{aligned} \tag{12.16}$$

The $kT(\mathbf{r}^{\alpha}_{\nu}, t)$ arises from the integration over momentum space. Note that the temperature $kT(\mathbf{r}^{\alpha}_{\nu}, t)$ depends only on the single position vector $\mathbf{r}^{\alpha}_{\nu}$, but $\bar{\Psi}_{\alpha}$, and therefore also $\mathbf{F}^{(b)\alpha}_{\nu}$, is a function of the $\mathbf{r}^{\alpha}_{\mu}$ for all μ. The second line of Eq. (12.16) follows from Eq. (12.3), if terms quadratic in $\mathbf{a}$ are neglected.

In a system which is at uniform temperature T throughout, Eq. (12.16) simplifies to the well-known expression:

$$\mathbf{F}^{(b)\alpha}_{\nu}(\mathbf{r}^{\alpha}, t) = -kT\frac{\partial}{\partial \mathbf{r}^{\alpha}_{\nu}}\ln \bar{\Psi}_{\alpha} \quad \text{Isothermal system} \tag{12.17}$$

This is the usual expression encountered in polymer kinetic theory discussions; however, for systems with temperature gradients, such as occur in the study of thermal conduction, Eq. (12.16) must be used instead.

We conclude this subsection by giving some double-bracket averages evaluated using the distribution function of Eq. (12.9), since these are needed later. To get the Brownian force expression in Eq. (12.16) we already have used:

$$[[(\dot{\mathbf{r}}_\nu^\alpha - \mathbf{u}_\nu^\alpha)(\dot{\mathbf{r}}_\mu^\alpha - \mathbf{u}_\mu^\alpha)]]^\alpha = (kT(\mathbf{r}_\nu^\alpha, t)/m_\nu^\alpha)\delta_{\nu\mu}\boldsymbol{\delta}$$
$$= (kT/m_\nu^\alpha)(1 + \mathbf{a}\cdot\mathbf{R}_\nu^\alpha)\delta_{\nu\mu}\boldsymbol{\delta} \quad (12.18)$$

in which $\boldsymbol{\delta}$ is the unit tensor, and the temperature T is a function of $\mathbf{r}$ and t. We note also that

$$[[(\dot{\mathbf{r}}_\nu^\alpha - \mathbf{u}_\nu^\alpha)\cdot(\dot{\mathbf{r}}_\nu^\alpha - \mathbf{u}_\nu^\alpha)]]^\alpha = (3kT/m_\nu^\alpha)(1 + \mathbf{a}\cdot\mathbf{R}_\nu^\alpha) \quad (12.19)$$

In Sect. 14 we will need the double-bracket quantity that appears in Eq. (7.8):

$$[[(\dot{\mathbf{r}}_\nu^\alpha - \mathbf{v})(\dot{\mathbf{r}}_\nu^\alpha - \mathbf{v})]]^\alpha = [[(\dot{\mathbf{r}}_\nu^\alpha - \mathbf{u}_\nu^\alpha)(\dot{\mathbf{r}}_\nu^\alpha - \mathbf{u}_\nu^\alpha)]]^\alpha + (\mathbf{u}_\nu^\alpha - \mathbf{v})(\mathbf{u}_\nu^\alpha - \mathbf{v})$$
$$= (kT/m_\nu^\alpha)(1 + \mathbf{a}\cdot\mathbf{R}_\nu^\alpha)\boldsymbol{\delta} + (\mathbf{u}_\nu^\alpha - \mathbf{v})(\mathbf{u}_\nu^\alpha - \mathbf{v}) \quad (12.20)$$

The cross terms drop out because the integrands are odd functions of $\dot{\mathbf{r}}_\nu^\alpha - \mathbf{u}_\nu^\alpha$. In Sect. 16 we need the double-bracket quantity in the first line of Eq. (8.11):

$$[[(\dot{\mathbf{r}}_\nu^\alpha - \mathbf{v})(\dot{\mathbf{r}}_\nu^\alpha - \mathbf{v})\cdot(\dot{\mathbf{r}}_\nu^\alpha - \mathbf{v})]]^\alpha = 2[[(\dot{\mathbf{r}}_\nu^\alpha - \mathbf{u}_\nu^\alpha)(\dot{\mathbf{r}}_\nu^\alpha - \mathbf{u}_\nu^\alpha)]]^\alpha\cdot(\mathbf{u}_\nu^\alpha - \mathbf{v})$$
$$+ [[(\dot{\mathbf{r}}_\nu^\alpha - \mathbf{u}_\nu^\alpha)\cdot(\dot{\mathbf{r}}_\nu^\alpha - \mathbf{u}_\nu^\alpha)]]^\alpha(\mathbf{u}_\nu^\alpha - \mathbf{v}) + (\mathbf{u}_\nu^\alpha - \mathbf{v})(\mathbf{u}_\nu^\alpha - \mathbf{v})\cdot(\mathbf{u}_\nu^\alpha - \mathbf{v})$$
$$= (5kT/m_\nu^\alpha)(1 + \mathbf{a}\cdot\mathbf{R}_\nu^\alpha)(\mathbf{u}_\nu^\alpha - \mathbf{v}) + (\mathbf{u}_\nu^\alpha - \mathbf{v})(\mathbf{u}_\nu^\alpha - \mathbf{v})\cdot(\mathbf{u}_\nu^\alpha - \mathbf{v}) \quad (12.21)$$

Here Eqs. (12.12) and (12.13) have been used, and four cross terms have dropped out because the integrands are odd functions. The factors $(\mathbf{u}_\nu^\alpha - \mathbf{v})$ appearing in the last two equations can be evaluated using Eq. (12.7).

Postulates other than Eq. (12.9) for the distribution in momentum space have been used. For example, some kinetic theorists have used a Maxwellian distribution about the velocity $\mathbf{v}$ at the center of mass of the molecule. Another possible assumption is that of a skewed distribution, in which additional empirical parameters are introduced so that the smoothed Brownian motion force may be stronger in the chain backbone direction than in the transverse directions [9, 20a, 20b, 20c, 21], [DPL, Sect. 13.7]. This idea has been proposed for describing the restricted motion of polymer chains in concentrated polymer solutions and in undiluted polymers. An extreme case of this is the "reptation assumption" [9, 14a], in which there is no Brownian force at all in the transverse directions, and the polymer chain is required on the average to slither back and forth along its backbone (DPL, Sect. 19.2b).

12.6 Abbreviations for Indicating Assumptions

In the next four sections, when we start doing detailed developments, the above assumptions – and some additional ones – will be introduced at various points. In order to keep track of the restrictions placed on various results, we include a set of letters just before many equations. These letters will indicate the assumptions that apply to the equation in question. The symbols used are as follows:

A	acceleration terms neglected – Eq. (12.5)
E	equilibration in momentum space Eq. (12.9)
F	scalar friction coefficient used to describe the hydrodynamic force (no hydrodynamic interaction) – Eq. (12.6)
S	short-range force approximation – Eq. (12.1)
C_m	concentration gradient constant over molecular dimensions
T_m	temperature gradient constant over molecular dimensions
V_m	velocity gradient constant over molecular dimensions
C	concentration gradient constant globally
T	temperature gradient constant globally
V	velocity gradient constant globally
T_c	temperature gradient independent of space and time
C_0	concentration gradient zero (that is, $\mathbf{b}_\alpha = 0$)
T_0	temperature gradient zero (that is, $\mathbf{a} = 0$)
V_0	velocity gradient zero (that is, $\boldsymbol{\kappa} = 0$)
D	dumbbell (Hookean spring, both bead masses and friction coefficients the same, no hydrodynamic interaction)
I	incompressible fluid
M	all bead masses identical within one molecule; all bead friction coefficients identical within one molecule
N	neglect of molecule-molecule interactions, except for solvent-solvent interactions
P	products and higher derivatives of $\mathbf{a}$ and $\mathbf{b}_\alpha$ are neglected
R	Rouse chain (Hookean springs, all bead masses and all friction coefficients the same, and no hydrodynamic interaction)
X	external forces omitted

With these symbols displayed by an equation, the reader will be spared the sometimes tedious task of deciding what restrictions have to be placed on the key results.

13 The "Diffusion Equation" for the Singlet Configuration-Space Distribution Function

In order to use the expressions for the mass flux vector, the stress tensor, and the energy flux vector in Table 1, it is necessary to know the singlet and doublet configuration-space distribution functions. For example, we need to solve Eq. (10.6) or Eq. (10.7) to get the distribution function for a single polymer molecule. This cannot, however, be done until something is inserted for the double-bracket quantities indicating momentum-space averages. Up to now there is not sufficient information about the phase-space distribution function f_α to allow for the calculation of these double-bracket quantities [16, 19]. We therefore freely make use of the friction coefficient empricism in Eqs. (12.6) and (12.7):

When Eqs. (12.7) and (12.10) are combined, we get: ⟨AEFT$_m$V$_m$⟩

$$\mathbf{u}_\nu^\alpha(\mathbf{r}^\alpha, t) - \mathbf{v}(\mathbf{r}_c^\alpha) = [\boldsymbol{\kappa} \cdot \mathbf{R}_\nu^\alpha] + \frac{1}{\zeta_\nu^\alpha}\left[- kT(1 + \mathbf{a} \cdot \mathbf{R}_\nu^\alpha)\frac{\partial}{\partial \mathbf{r}_\nu^\alpha} \ln \bar{\Psi}_\alpha - kT\mathbf{a} + \mathbf{F}_\nu^{(\phi)\alpha} + \mathbf{F}_\nu^{(e)\alpha} \right] \tag{13.1}$$

Here, we have assumed that the temperature is a linear function of position (see Eq. (12.3)). When this expression for $\mathbf{u}_\nu^\alpha = [[\dot{\mathbf{r}}_\nu^\alpha]]^\alpha$ is inserted into Eq. (10.6), an equation for the configurational distribution function $\bar{\Psi}_\alpha$ is obtained. The resulting equation is often referred to as the *diffusion equation for* $\bar{\Psi}_\alpha$.

Often, however, we prefer to work with the center-of-mass vector $\mathbf{r}_c^\alpha$ and the connector vectors $\mathbf{Q}_k^\alpha$. By using Eqs. (2.3) and (2.13), we can obtain from Eq. (13.1) the following two equations (for models in which all bead masses, m, and friction coefficients, ζ, are identical): ⟨AEFMT$_m$V$_m$⟩

$$[[\dot{\mathbf{r}}_c^\alpha]] = \mathbf{v}(\mathbf{r}_c^\alpha) - \frac{kT}{\zeta}\left[\frac{1}{N_\alpha}\frac{\partial}{\partial \mathbf{r}_c^\alpha}\ln \Psi_\alpha + \frac{1}{N_\alpha}\sum\nolimits_k (\mathbf{a} \cdot \mathbf{Q}_k^\alpha)\frac{\partial}{\partial \mathbf{Q}_k^\alpha}\ln \Psi_\alpha + \mathbf{a}\right] + \frac{1}{N_\alpha \zeta}\mathbf{F}^{(e)\alpha} \tag{13.2}$$

$$[[\dot{\mathbf{Q}}_j^\alpha]]^\alpha = [\boldsymbol{\kappa} \cdot \mathbf{Q}_j^\alpha] - \frac{kT}{\zeta}\left[\sum_k A_{jk}\left[\frac{\partial \ln \Psi_\alpha}{\partial \mathbf{Q}_k^\alpha} + \frac{1}{kT}\frac{\partial \phi^\alpha}{\partial \mathbf{Q}_k^\alpha}\right] + (\mathbf{a} \cdot \mathbf{Q}_j^\alpha)\left(\frac{1}{N_\alpha}\frac{\partial}{\partial \mathbf{r}_c^\alpha}\ln \Psi_\alpha\right) + \left(\mathbf{a} \cdot \sum\nolimits_k \sum\nolimits_l D_{jkl}\mathbf{Q}_l^\alpha \frac{\partial}{\partial \mathbf{Q}_k^\alpha}\ln \Psi_\alpha\right)\right] + \frac{1}{\zeta}\sum\nolimits_\nu \bar{B}_{j\nu}\mathbf{F}_\nu^{(e)\alpha} \tag{13.3}$$

In which the $A_{jk} = \Sigma_\nu \bar{B}_{j\nu}\bar{B}_{k\nu}$ are the elements of the Rouse matrix (see Eq. (2.12). The $D_{jkl} = \Sigma_\nu \bar{B}_{j\nu}\bar{B}_{k\nu}B_{\nu l}$ are the elements of a matrix which, as far as we know, has never been discussed. All these matrices should carry an index α, but we omit this index for brevity. For $N_\alpha = 2$ (elastic dumbbells) there is just one element, D_{111}, which is zero. For $N_\alpha = 3$ we have:

$$D_{111} = -\tfrac{1}{3};\quad D_{112} = -\tfrac{2}{3};\quad D_{121} = -\tfrac{1}{3};\quad D_{122} = +\tfrac{1}{3}$$

$$D_{211} = -\tfrac{1}{3};\quad D_{212} = +\tfrac{1}{3};\quad D_{221} = +\tfrac{2}{3};\quad D_{222} = +\tfrac{1}{3}$$

When the double-bracketed quantities of Eqs. (13.2) and (13.3) are substituted into Eq. (10.7), the partial differential equation (the "diffusion equation") for the distribution function Ψ_α results. In principle this equation can now be solved, as soon as the velocity and temperature fields have been specified by giving $\boldsymbol{\kappa}$ and $\mathbf{a}$, and the polymer model has been specified by giving the potential energies needed to describe the intramolecular and external forces.

The terms containing $\mathbf{a} = \nabla \ln T$ Eqs. (13.1), (13.2) and (13.3) do not seem to have been presented before; they may be omitted when isothermal systems are being considered. According to Eq. (13.2) there may be a "drift" of the center of mass associated with thermal gradients – that is, a *thermal diffusion* effect. This effect is discussed further in Sects. 15 and C.3.

13.1 Fluids with Constant Temperature and Constant Composition

When Eqs. (13.2) and (13.3) are inserted into Eq. (10.7) and external forces are neglected, we get the following "diffusion" equation for Ψ_α for incompressible systems at constant temperature: $\langle \mathrm{AEFIMXT_0V_m} \rangle$

$$\frac{\partial}{\partial t}\Psi_\alpha + (\mathbf{v}\cdot\nabla\Psi_\alpha) = -\frac{kT}{N_\alpha\zeta}(\nabla\cdot\nabla\Psi_\alpha) - \sum_j \frac{\partial}{\partial \mathbf{Q}_j^\alpha}\cdot[\boldsymbol{\kappa}\cdot\mathbf{Q}_j^\alpha]\Psi_\alpha + \frac{kT}{\zeta}\sum_j\sum_k A_{jk}\left(\frac{\partial}{\partial \mathbf{Q}_j^\alpha}\cdot\left(\frac{\partial\Psi_\alpha}{\partial \mathbf{Q}_k^\alpha} + \frac{\Psi_\alpha}{kT}\frac{\partial\phi^\alpha}{\partial \mathbf{Q}_k^\alpha}\right)\right) \tag{13.4}$$

The number of exact analytical solutions of Eq. (13.4) for special flows is quite small. We present here the solutions given by Lodge and Wu [21] and by van Wiechen and Booij [22] for solutions of Rouse chains for arbitrary, time-dependent flows. We also give the results for Hookean dumbbells. These models are known not to be very satisfactory, their most glaring shortcomings being that they do not give viscosity and normal stress coefficients that decrease with shear rate, and that the elongational viscosity becomes infinite at a finite elongation rate. Nonetheless they have been useful for exploratory kinetic theory calculations, because analytical solutions can be obtained for many kinds of problems. We will, in fact, use these models in the next three sections for illustrative purposes.

13.1.1 Dilute Solutions of Hookean Dumbbells at Constant Temperature

This model consists of two identical beads with bead friction coefficient ζ joined by a Hookean spring with spring constant H.

For this model, there is but one connector vector, called $\mathbf{Q}$ (the vector from bead "1" to bead "2") and only one term in the summations in Eq. (13.4). There is only one element in the Rouse matrix (see Eq. (2.12)), namely $A_{11} = 2$, and $\phi^\alpha = \frac{1}{2}H(\mathbf{Q}\cdot\mathbf{Q})$. For this system we postulate as a solution of Eq. (13.4) a normalized Gaussian distribution of the form:

$$\Psi_\alpha(\mathbf{r},\mathbf{Q},t) = n_\alpha(\mathbf{r},t)\,\psi_\alpha(\mathbf{Q},t) = \frac{n_\alpha(H/2\pi kT)^{3/2}}{\sqrt{\det\boldsymbol{\alpha}}}\,e^{-(H/2kT)(\boldsymbol{\alpha}^{-1}:\mathbf{QQ})} \tag{13.5}$$

in which the tensor $\boldsymbol{\alpha}$ is a function of $\mathbf{r}$ and t. When this postulated form for the solution is substituted into Eq. (13.4) with the $\nabla^2\Psi_\alpha$ term omitted (since it is negligible when T, n_α, and $\boldsymbol{\kappa}$ are constant over molecular dimensions), the following equation for α is found: $\langle \mathrm{ADEFIXC_mT_0V_m} \rangle$

$$\boldsymbol{\alpha} + \lambda_H \boldsymbol{\alpha}_{(1)} = \boldsymbol{\delta} \tag{13.6}$$

where $\boldsymbol{\delta}$ is the unit tensor. The initial condition for Eq. (13.6) is taken to be $\boldsymbol{\alpha} = \boldsymbol{\delta}$ at $t = -\infty$. Here $\boldsymbol{\alpha}_{(1)}$ is a *convected time derivative*, defined by $\boldsymbol{\alpha}_{(1)} = D\boldsymbol{\alpha}/Dt - \{\boldsymbol{\kappa}\cdot\boldsymbol{\alpha} + \boldsymbol{\alpha}\cdot\boldsymbol{\kappa}^\dagger\}$, where $D\boldsymbol{\alpha}/Dt = \partial\boldsymbol{\alpha}/\partial t + \{\mathbf{v}\cdot\nabla\boldsymbol{\alpha}\}$ is the *substantial derivative* (see Sect. D.2 of DPL), and $\lambda_H = \zeta/4H$ is the characteristic time for Hookean dumbbells. When $\boldsymbol{\kappa}(t)$ is independent of position and $\mathbf{b}_\alpha(\mathbf{r}, t)$ is independent of position over molecular dimensions (cf. Eqs. (15.3-22) and (20.2-14) of DPL), the solution to Eq. (13.6) is: $\langle \mathrm{ADEFIXC_mT_0V} \rangle$

$$\begin{aligned} \boldsymbol{\alpha} &= \frac{1}{\lambda_H}\int_{-\infty}^{t} e^{-(t-t')/\lambda_H}\mathbf{B}(t,t')dt' \\ &= \boldsymbol{\delta} - \frac{1}{\lambda_H}\int_{-\infty}^{t} e^{-(t-t')/\lambda_H}\boldsymbol{\gamma}_{[0]}(t,t')dt' \equiv \boldsymbol{\delta} - \boldsymbol{\Gamma} \end{aligned} \tag{13.7}$$

Here $\mathbf{B}$ is the standard *Finger strain tensor* used in continuum mechanics, and $\boldsymbol{\gamma}_{[0]} = \boldsymbol{\delta} - \mathbf{B}$ is a *relative finite strain tensor*, defined in DPL, Eq. D.3-4. We note in passing that it follows from Eq. (13.5) that the quantity $(H/kT)\int \mathbf{QQ}\Psi_\alpha\, d\mathbf{Q}$ is equal to $\boldsymbol{\alpha}$ and thus satisfies Eq. (13.6).

13.1.2 Dilute Solutions of Rouse Chains at Constant Temperature

This model consists of N_α identical beads each with friction coefficient ζ joined together linearly by Hookean springs each with spring constant H. For this model the solution to Eq. (13.4) is:

$$\begin{aligned} \Psi_\alpha(\mathbf{r}, \mathbf{Q}^\alpha, t) &= n_\alpha(\mathbf{r}, t)\psi_\alpha(\mathbf{Q}^\alpha, t) \\ &= \frac{n_\alpha(H/2\pi kT)^{3(N_\alpha-1)/2}}{\Pi_j\sqrt{\det\boldsymbol{\alpha}_j}}\, e^{-(H/2kT)\Sigma_j(\boldsymbol{\alpha}_j^{-1}:\mathbf{Q}_j'\mathbf{Q}_j')} \end{aligned} \tag{13.8}$$

in which the $\boldsymbol{\alpha}_j$ are the solutions to the equations: $\langle \mathrm{AEFIRXC_mT_0V_m} \rangle$

$$\boldsymbol{\alpha}_j + \lambda_j\boldsymbol{\alpha}_{j(1)} = \boldsymbol{\delta} \quad \text{for } j = 1, 2, \ldots N_\alpha - 1 \tag{13.9}$$

with the initial conditions $\boldsymbol{\alpha}_j = \boldsymbol{\delta}$ for all j. The solutions are: $\langle \mathrm{AEFIRXC_mT_0V} \rangle$

$$\begin{aligned} \boldsymbol{\alpha}_j &= \frac{1}{\lambda_j}\int_{-\infty}^{t} e^{-(t-t')/\lambda_j}\mathbf{B}(t,t')dt' \\ &= \boldsymbol{\delta} - \frac{1}{\lambda_j}\int_{-\infty}^{t} e^{-(t-t')/\lambda_j}\boldsymbol{\gamma}_{[0]}(t,t')dt' \equiv \boldsymbol{\delta} - \boldsymbol{\Gamma}_j \end{aligned} \tag{13.10}$$

Here the $\lambda_j = \zeta c_j/2H$ are the time constants for the N_α-bead Rouse model, with the c_j being the eigenvalues of the Kramers matrix $C_{ij} = \Sigma_\nu B_{\nu i} B_{\nu j}$, which is the inverse of the Rouse matrix $A_{ij} = \Sigma_\nu \bar{B}_{i\nu} \bar{B}_{j\nu}$ (see Eq. (2.12)). The primed "normal coordinates" in Eq. (13.8) are related to the internal coordinates defined in Eq. (2.2) by:

$$\mathbf{Q}_j = \Sigma_k \Omega_{jk} \mathbf{Q}'_k \quad \text{and} \quad \frac{\partial}{\partial \mathbf{Q}_j} = \Sigma_k \Omega_{jk} \frac{\partial}{\partial \mathbf{Q}'_k} \tag{13.11}$$

where $\Omega_{jk} = \sqrt{2/N_\alpha}\sin(jk\pi/N_\alpha)$, for which

$$\Sigma_i \Omega_{ij} \Omega_{ik} = \delta_{jk} \quad \text{and} \quad \Sigma_j \Sigma_k \Omega_{ji} C_{jk} \Omega_{kl} = c_i \delta_{il} \tag{13.12}$$

That is, the Ω_{jk} are the elements of an $(N_\alpha - 1) \times (N_\alpha - 1)$ orthogonal matrix that diagonalizes the Kramers and Rouse matrices. For the complete derivation of the above results, see DPL Sect. 15.3.

It must be kept in mind that the tensor $\mathbf{B}(t, t')$ is a functional of the tensor $\boldsymbol{\kappa}$ (see, for example, DPL, Problem 9D.1) in Eqs. (13.7) and (13.10). In deriving these equations, we have made the assumption that the flow is homogeneous, that is, $\boldsymbol{\kappa}$ is independent of position but not of time. Since the integration on t' in these equations is an integration following a fluid element, weighted by $\exp[-(t-t')/\lambda_j]$, it is seen that $\boldsymbol{\kappa}$ must be known for a period of time large compared to the longest λ_j. Because during this time interval the fluid element may travel a distance considerably larger than molecular dimensions, $\boldsymbol{\kappa}$ must be known over this distance. For this reason one must question the validity of the approximation—which is frequently made—that expressions derived assuming homogeneous flow may be applied to inhomogeneous flow fields (see also comments at the end of Sect. 14.1).

Other solutions for the singlet configuration-space distribution function are those for the steady-state, homogeneous potential flow of elastic dumbbells with any kind of spring (DPL, Eq. (13.2–14)), and the first few terms in a perturbation solution for steady-state, homogeneous flow of FENE dumbbells (DPL, Eq. (13.2–15)).

Brownian and molecular dynamics simulations are promising as methods for expanding our knowledge about chain motions and distribution functions. For example, in Brownian dynamics simulations for large molecules, represented by, say, 100 beads, direct solution of the Langevin equations corresponding to the above diffusion equation is numerically more feasible than solving the diffusion equations with many independent variables. For a discussion of the solution of the Langevin equations by stochastic methods, see the recently published book of Öttinger [23].

13.2 Fluids with Varying Temperature and Concentration Gradients

We begin by defining a function $\psi_\alpha(\mathbf{r}, \mathbf{Q}^\alpha, t)$ by the relation:

$$\Psi_\alpha(\mathbf{r}, \mathbf{Q}^\alpha, t) = n_\alpha(\mathbf{r}, t)\, \psi_\alpha(\mathbf{r}, \mathbf{Q}^\alpha, t) \tag{13.13}$$

where the species number density n_α is given by Eq. (4.5). We now use this definition, along with the expressions for the temperature and composition given in Eqs. (12.3) and (12.4), in the expression for $[[\dot{\mathbf{r}}_\nu^\alpha]]^\alpha$ in Eq. (12.7). We retain terms containing $\mathbf{a}$, $\mathbf{b}_\alpha$, and $\nabla\psi_\alpha$ but neglect terms quadratic in these quantities. We note that according to the diffusion equation used in (a), ψ_α is independent of $\mathbf{r}$ and hence $\nabla\psi_\alpha$ is zero. Accordingly, if $\mathbf{a}$ and $\mathbf{b}_\alpha$ are small, one can anticipate that $\nabla\psi_\alpha$ will be small also (we note parenthetically that $\nabla\psi_\alpha$ is proportional to $\mathbf{a}$ for a Rouse chain). Then we get the following expression for the averaged bead velocity: ⟨AEFMC$_m$T$_m$V$_m$⟩

$$\begin{aligned}[][\dot{\mathbf{r}}_\nu^\alpha]] \equiv \mathbf{u}_\nu^\alpha(\mathbf{r}_c, \mathbf{Q}^\alpha, t) = \mathbf{v}(\mathbf{r}_c^\alpha) + [\boldsymbol{\kappa}\cdot\Sigma_k B_{\nu k}\mathbf{Q}_k^\alpha] \\ -\frac{kT}{\zeta}\left(\mathbf{a} + \frac{1}{N_\alpha}\mathbf{b}_\alpha + \frac{1}{N_\alpha}\frac{\partial}{\partial\mathbf{r}_c^\alpha}\ln\psi_\alpha + \sum_k \bar{B}_{k\nu}\frac{\partial}{\partial\mathbf{Q}_k^\alpha}\ln\psi_\alpha\right. \\ \left.+\left(\mathbf{a}\cdot\sum_j\sum_k \bar{B}_{k\nu}B_{\nu j}\mathbf{Q}_j^\alpha\frac{\partial}{\partial\mathbf{Q}_k^\alpha}\ln\psi_\alpha\right)\right) \\ -\frac{1}{\zeta}\sum_k \bar{B}_{k\nu}\frac{\partial\phi^\alpha}{\partial\mathbf{Q}_k^\alpha} + \frac{1}{\zeta}\mathbf{F}_\nu^{(e)\alpha}\end{aligned} \tag{13.14}$$

This is now substituted into Eq. (10.6), which then becomes (when the external forces are neglected): ⟨AEFMPXC$_m$T$_m$V$_m$⟩

$$\begin{aligned}\frac{\partial}{\partial t}n_\alpha\psi_\alpha = {} & -n_\alpha\psi_\alpha(\nabla\cdot\mathbf{v}) - n_\alpha(\mathbf{v}\cdot(\nabla\psi_\alpha + \mathbf{b}_\alpha\psi_\alpha)) + \frac{kTn_\alpha}{N_\alpha\zeta}\nabla^2\psi_\alpha \\ & - n_\alpha\sum_j\frac{\partial}{\partial\mathbf{Q}_j^\alpha}\cdot[\boldsymbol{\kappa}\cdot\mathbf{Q}_j^\alpha]\psi_\alpha + \frac{kTn_\alpha}{\zeta}\sum_j\sum_k A_{jk}\left(\frac{\partial}{\partial\mathbf{Q}_j^\alpha}\cdot\left(\frac{\partial\psi_\alpha}{\partial\mathbf{Q}_k^\alpha} + \frac{\psi_\alpha}{kT}\frac{\partial\phi^\alpha}{\partial\mathbf{Q}_k^\alpha}\right)\right) \\ & + \frac{kTn_\alpha}{\zeta}\sum_j\sum_k\sum_l\left(\frac{\partial}{\partial\mathbf{Q}_j^\alpha}\cdot\left[\mathbf{a}\cdot D_{jkl}\mathbf{Q}_l^\alpha\frac{\partial}{\partial\mathbf{Q}_k^\alpha}\psi_\alpha\right]\right)\end{aligned} \tag{13.15}$$

in which the D_{jkl} are matrix elements defined just after Eq. (13.3). Equation (2.13) was used to convert the derivatives with respect to bead coordinates into derivatives with respect to the center of mass and the connector vectors. The time differentiation of the product on the left side is now performed and then use is made of the following version of Eq. (6.1):

$$\begin{aligned}\frac{\partial n_\alpha}{\partial t} &= -(\nabla\cdot n_\alpha\mathbf{v}) - \frac{1}{m_m^\alpha}(\nabla\cdot\mathbf{j}_\alpha) \\ &= -n_\alpha(\nabla\cdot\mathbf{v}) - (\mathbf{v}\cdot\nabla n_\alpha) + \left(\nabla\cdot\frac{n_\alpha kT}{N_\alpha\zeta}(\mathbf{b}_\alpha + \mathbf{a})\right)\end{aligned} \tag{13.16}$$

in which we have made use of Eq. (15.6) for the mass flux (with no external forces). When use is made of this result, and when in addition the $\nabla^2\psi_\alpha$ term is neglected (because it is of higher order), we get finally for the "diffusion equation" for the distribution function $\psi_\alpha(\mathbf{r}, \mathbf{Q}^\alpha, t)$: ⟨AEFMPXC$_mT_mV_m$⟩

$$\frac{D}{Dt}\psi_\alpha \equiv \frac{\partial}{\partial t}\psi_\alpha + (\mathbf{v}\cdot\nabla\psi_\alpha) = -\sum_j \frac{\partial}{\partial \mathbf{Q}_j^\alpha}\cdot[\kappa\cdot\mathbf{Q}_j^\alpha]\psi_\alpha$$
$$+\frac{kT}{\zeta}\sum_j\sum_k A_{jk}\left(\frac{\partial}{\partial \mathbf{Q}_j^\alpha}\cdot\left(\frac{\partial\psi_\alpha}{\partial \mathbf{Q}_k^\alpha} + \frac{\psi_\alpha}{kT}\frac{\partial\phi^\alpha}{\partial \mathbf{Q}_k^\alpha}\right)\right)$$
$$+\frac{kT}{\zeta}\sum_j\sum_k\sum_l\left(\frac{\partial}{\partial \mathbf{Q}_j^\alpha}\cdot\left[\mathbf{a}\cdot D_{jkl}\mathbf{Q}_l^\alpha\frac{\partial}{\partial \mathbf{Q}_k^\alpha}\psi_\alpha\right]\right) \tag{13.17}$$

It should be observed that in this equation the vector $\mathbf{a} = \nabla \ln T$ appears, but the vector $\mathbf{b}_\alpha = \nabla \ln n_\alpha$ has dropped out. Equation (13.17) is valid for any bead-spring model for which all beads and friction coefficients are the same; also external forces have been neglected in developing this result.

We now give the solution to Eq. (13.17) for the simplest possible model, namely Hookean dumbbells, for which there is but one element in the D-matrix and it is identically equal to zero. For this equation we can postulate a Gaussian-form solution:

$$\psi_\alpha(\mathbf{r}, \mathbf{Q}, t) = \frac{(H/2\pi kT)^{3/2}}{\sqrt{\det\boldsymbol{\alpha}}}\, e^{-(H/2kT)(\boldsymbol{\alpha}^{-1}:\mathbf{QQ})}$$
$$= \pi^{3/2}\sqrt{\det(H/2kT)\boldsymbol{\alpha}^{-1}}\, e^{-(H/2kT)(\boldsymbol{\alpha}^{-1}:\mathbf{QQ})} \tag{13.18}$$

in which α and T are now functions of $\mathbf{r}$ and t. Since $\boldsymbol{\alpha}$ and T always appear in the combination $\boldsymbol{\alpha}T$, this product should be the solution to the convected differential equation (cf Eq. (13.6)): ⟨ADEFPXC$_m$T$_m$V$_m$⟩

$$\boldsymbol{\alpha}T + \lambda_H(\boldsymbol{\alpha}T)_{(1)} = \boldsymbol{\delta}T \tag{13.19}$$

which can be solved by the same method used to solve Eq. (13.6); the solution is ⟨ADEFPXC$_m$T$_m$V⟩

$$\boldsymbol{\alpha}(\mathbf{r}, t) = \frac{1}{\lambda_H}\int_{-\infty}^{t} e^{-(t-t')/\lambda_H}\left(\frac{T(\mathbf{r}, t, t')}{T(\mathbf{r}, t)}\right)\mathbf{B}(t, t')dt' \tag{13.20}$$

in which $\mathbf{B}(t,t')$ is the Finger strain tensor introduced in Eq. (13.7) for the homogeneous flow, and $T(\mathbf{r},t,t')$ is the temperature of the fluid particle $(\mathbf{r}, t)$ at the time t'. For isothermal flows this result simplifies to that in Eq. (13.7). If we now make use of the linear-gradient assumption in Eq. (12.3), and if we further assume that $\mathbf{a}$ is independent of time, then Eq. (13.20) may be written as: ⟨ADEFPXC$_m$T$_c$V⟩

$$\boldsymbol{\alpha}(\mathbf{r}, t) = \boldsymbol{\alpha}^{(0)}(t) + \boldsymbol{\alpha}^{(1)}(\mathbf{r}, t) = \frac{1}{\lambda_H}\int_{-\infty}^{t} e^{-(t-t')/\lambda_H}\mathbf{B}(t, t')dt'$$
$$-\left(\mathbf{a}\cdot\frac{1}{\lambda_H}\int_{-\infty}^{t} e^{-(t-t')/\lambda_H}(\mathbf{r}-\mathbf{r}')\mathbf{B}(t, t')dt'\right) \tag{13.21}$$

in which $\mathbf{a} = \nabla \ln T$, and $(\mathbf{r} - \mathbf{r}')$ is the displacement of the fluid particle $(\mathbf{r}, t)$ between times t' and t. This distance is easily calculated for simple flows, but it is not possible to give a general formal expression for it (DPL Sect. 8.1). Note that, because of the appearance of $(\mathbf{r} - \mathbf{r}')$ in the expression for $\boldsymbol{\alpha}^{(1)}$, the distribution function $\psi_\alpha(\mathbf{r}, \mathbf{Q}, t)$ depends explicitly on the velocity of the fluid, that is, the velocity in a frame in which $\mathbf{a}$ does not depend on t. Equation (13.21) is used in Sect. 16 for deriving an expression for the thermal conductivity tensor of Hookean dumbbell solutions.

It should be noted that we have taken the friction coefficient ζ and the spring constant H to be independent of the temperature, and therefore the time constant λ_H is temperature independent. Although it might be appropriate to take into account the variation of the time constant with temperature, it does not seem to be possible to do so in the present formulation of the kinetic theory, because of the assumption made in Eq. (2.24) that the intramolecular potential depends only on the interbead distances and not on the position in space or on the time. Allowing H to be a function of $T(\mathbf{r},t)$ would be in violation of the assumption in Eq. (2.24). Numerous papers have appeared in the literature in which H is taken to be temperature-dependent, and we question the correctness of the results.

For flows in which fluid particles move rectilinearly, the second term in Eq. (13.21) vanishes if the temperature gradient is perpendicular to the direction of flow. Therefore one can anticipate that there may be substantial differences in the behavior of the fluxes depending on the relative directions of the flow and the direction of the temperature gradient.

13.3 The Tensor $\alpha = \alpha^{(0)} + \alpha^{(1)}$ for Hookean Dumbbells for Steady-State Shear and Elongational Flows

In subsequent sections we will need the expressions for the $\boldsymbol{\alpha}$ tensor of Eq. (13.21) for steady shear flow and steady elongational flow.

For *steady shear flow*, with $v_x = \dot{\gamma} y$, $v_y = 0$, and $v_z = 0$, the nonzero components of the two parts of the $\boldsymbol{\alpha}$ tensor are given by:

$$\alpha_{xx}^{(0)} = 1 + 2\lambda_H^2 \dot{\gamma}^2; \quad \alpha_{yy}^{(0)} = \alpha_{zz}^{(0)} = 1; \quad \alpha_{xy}^{(0)} = \alpha_{yx}^{(0)} = \lambda_H \dot{\gamma} \tag{13.22}$$

$$\alpha_{xx}^{(1)} = -a_x y \lambda_H \dot{\gamma}(1 + 6\lambda_H^2 \dot{\gamma}^2); \quad \alpha_{yy}^{(1)} = \alpha_{zz}^{(1)} = -a_x y \lambda_H \dot{\gamma};$$

$$\alpha_{xy}^{(1)} = \alpha_{yx}^{(1)} = -2a_x y \lambda_H^2 \dot{\gamma}^2 \tag{13.23}$$

in which $\lambda_H = \zeta/4H$ is the time constant for the Hookean dumbbell, and $\dot{\gamma}$ is the constant "shear rate."

For *steady elongational flow*, with $v_x = -\frac{1}{2}\dot{\varepsilon}x$, $v_y = -\frac{1}{2}\dot{\varepsilon}y$, $v_z = \dot{\varepsilon}z$, only the diagonal elements of the $\boldsymbol{\alpha}$ tensor are nonzero, and they may be written as:

$$\alpha_{xx}^{(0)} = \alpha_{yy}^{(0)} = \frac{1}{1 + \lambda_H \dot{\varepsilon}}; \quad \alpha_{zz}^{(0)} = \frac{1}{1 - 2\lambda_H \dot{\varepsilon}} \tag{13.24}$$

$$\alpha_{xx}^{(1)} = \alpha_{yy}^{(1)} = \frac{\frac{1}{2}\lambda_H\dot{\varepsilon}(a_x x + a_y y)}{(1 + \lambda_H\dot{\varepsilon})(1 + \frac{1}{2}\lambda_H\dot{\varepsilon})} - \frac{\lambda_H\dot{\varepsilon}(a_z z)}{(1 + \lambda_H\dot{\varepsilon})(1 + 2\lambda_H\dot{\varepsilon})};$$

$$\alpha_{zz}^{(1)} = \frac{\frac{1}{2}\lambda_H\dot{\varepsilon}(a_x x + a_y y)}{(1 - 2\lambda_H\dot{\varepsilon})(1 - \frac{5}{2}\lambda_H\dot{\varepsilon})} - \frac{\lambda_H\dot{\varepsilon}(a_z z)}{(1 - 2\lambda_H\dot{\varepsilon})(1 - \lambda_H\dot{\varepsilon})} \tag{13.25}$$

in which $\dot{\varepsilon}$ is the constant "elongational rate". The infinities that arise in these expressions when the denominators go to zero are a manifestation of the physically unrealistic model with unlimited extensibility. The above expressions are obtained by using the tabulations for the finite strain tensors and the displacement functions in Table C.1 of Appendix C of DPL.

For the Rouse model the components of the $\boldsymbol{\alpha}_j$-tensors of Eq. 13.10, for isothermal flow, may be obtained from the above expressions for the components of the $\boldsymbol{\alpha}^{(0)}$-tensor by replacing λ_H by λ_j.

13.4 Some Integrals for the Hookean Dumbbell Model

Both Eq. (13.5) (isothermal systems) and Eq. (13.8) (nonisothermal systems) have the same form, ψ_α in both equations being given in terms of a symmetric tensor $\boldsymbol{\alpha}$. In the next several sections various integrals over $\mathbf{Q}$ involving the distribution function ψ_α arise, and hence we summarize them here:

$$\int \mathbf{Q}\mathbf{Q}\psi_\alpha d\mathbf{Q} = (kT/H)\boldsymbol{\alpha} \tag{13.26}$$

$$\int Q^2\psi_\alpha d\mathbf{Q} = (kT/H)(tr\boldsymbol{\alpha}) \tag{13.27}$$

$$\int \mathbf{Q}\mathbf{Q}\mathbf{Q}\mathbf{Q}\psi_\alpha d\mathbf{Q} = (kT/H)^2\, \mathbf{I}\boldsymbol{\alpha}\boldsymbol{\alpha}\mathbf{I} \tag{13.28}$$

$$\int Q^2\mathbf{Q}\mathbf{Q}\psi_\alpha d\mathbf{Q} = (kT/H)^2[\boldsymbol{\alpha}(tr\boldsymbol{\alpha}) + 2\{\boldsymbol{\alpha}\cdot\boldsymbol{\alpha}]\} \tag{13.29}$$

$$\boldsymbol{\alpha}^{-1}: \int \mathbf{Q}\mathbf{Q}\mathbf{Q}\mathbf{Q}\psi_\alpha d\mathbf{Q} = 5(kT/H)^2\boldsymbol{\alpha} \tag{13.30}$$

$$\boldsymbol{\alpha}^{-1}: \int Q^2\mathbf{Q}\mathbf{Q}\psi_\alpha d\mathbf{Q} = 5(kT/H)^2(tr\boldsymbol{\alpha}) \tag{13.31}$$

in which

$$\mathbf{I}\boldsymbol{\alpha}\boldsymbol{\alpha}\mathbf{I}_{mnpq} = \alpha_{mn}\alpha_{pq} + \alpha_{mp}\alpha_{nq} + \alpha_{mq}\alpha_{np} \tag{13.32}$$

are the components of the fully symmetrized fourth-order tensor formed from from the tensor $\boldsymbol{\alpha}\boldsymbol{\alpha}$.

14 The Stress Tensor and Rheological Constitutive Equations

Once the singlet distribution function Ψ_α has been found, we are in a position to evaluate the various contributions to the fluxes that depend on Ψ_α (see Table 1). In this section we discuss the contributions to the stress tensor, and in the next two sections the contributions to the mass and heat flux vectors. In these sections, for illustrative purposes, we restrict ourselves to the Rouse bead-spring chain and the Hookean dumbbell models, for which we can use the singlet distribution functions Ψ_α given in Eqs. (13.5) and (13.8).

We confine our attention here to dilute solutions of several polymer species in a solvent. According to Sect. 7, the stress tensor is a sum of four contributions, the first three of which involve the singlet distribution function(s), whereas the fourth involves the doublet distribution function(d):

$$\boldsymbol{\pi} = [\boldsymbol{\pi}^{(k)} + \boldsymbol{\pi}^{(\phi)} + \boldsymbol{\pi}^{(e)}] + \boldsymbol{\pi}^{(d)} = [\Sigma_\alpha \boldsymbol{\pi}_\alpha^{(s)}] + \Sigma_\alpha \Sigma_\beta \boldsymbol{\pi}_{\alpha\beta}^{(d)} \tag{14.1}$$

As shown here, we can also regard the stress tensor as being a sum of contributions over the individual chemical species plus a double sum over all pairs of species. If the solvent species (subscript "s") is now treated separately, and primes indicate sums over all solute species, then Eq. (14.1) can be written as:

$$\boldsymbol{\pi} = \boldsymbol{\pi}_s^{(s)} + \Sigma'_\alpha \boldsymbol{\pi}_\alpha^{(s)} + \boldsymbol{\pi}_{ss}^{(d)} + \Sigma'_\alpha \boldsymbol{\pi}_{s\alpha}^{(d)} + \Sigma'_\alpha \Sigma'_\beta \boldsymbol{\pi}_{\alpha\beta}^{(d)} \tag{14.2}$$

It has been common practice in the kinetic theory of dilute solutions to lump the first and third terms together and refer to this combination as $\boldsymbol{\pi}_s$, the solvent contribution, and to neglect the fourth and fifth terms. Omitting the fifth term is appropriate for dilute solutions in which polymer-polymer interactions can be ignored. Neglect of the fourth term, however, may not always be appropriate; omission of this term disregards solvent effects, such as the phenomenon of "solvent modification" that has been observed in recent years [23a, 23b, 23c] This term refers to the tendency for solvent molecules to become preferentially oriented in the neighborhood of the polymer molecules.

In this section we will then use the simplified version of Eq. (14.2) as has been customary in dilute solution theories [1,2,3,4] and write for a solution with a single solute α: ⟨N⟩

$$\boldsymbol{\pi} = \boldsymbol{\pi}_s + \boldsymbol{\pi}_\alpha \tag{14.3}$$

The solvent contribution $\boldsymbol{\pi}_s$ is then regarded as an experimentally determined quantity, and the kinetic theory is then used to obtain an expression for the solute contribution $\boldsymbol{\pi}_\alpha$. The latter is then regarded as the sum of a kinetic contribution $\boldsymbol{\pi}^{(k)}$ (from Eq. (7.8)) and an intermolecular force contribution $\boldsymbol{\pi}^{(\phi)}$ (from Eq. (7.15)), the external force contribution being omitted (except in Sect. 14.4).

14.1 The Stress Tensor for a Solution of Rouse Chains

It is necessary to introduce the notation that we will use here and in the next two sections regarding the successive terms that are produced when the distribution function and the momentum averages, which appear in the contributions to the stress tensor, are expanded in Taylor series. The first, second, third, terms resulting from the Taylor expansions of the fluxes (and not of the source terms!) are indicated thus: $\boldsymbol{\pi}_\alpha(1)$, $\boldsymbol{\pi}_\alpha(2)$, $\boldsymbol{\pi}_\alpha(3), \ldots$, the number in parentheses designating the "Taylor order."

14.1.1 First-Order Contribution $\boldsymbol{\pi}_\alpha(1)$ to the Stress Tensor

We begin with $\boldsymbol{\pi}^{(k)}(1)$, the kinetic contribution to the stress tensor at first order, given by the first term in the Taylor series in the second line of Eq. 7.8. In this equation we replace $\dot{\mathbf{r}}_\nu^\alpha - \mathbf{v}$ by $\dot{\mathbf{r}}_\nu^\alpha - \mathbf{u}_\nu^\alpha$ and then add a compensating term. This gives:

$$\boldsymbol{\pi}^{(k)}(1) = \boldsymbol{\pi}_1^{(k)}(1) + \boldsymbol{\pi}_2^{(k)}(1) \tag{14.4}$$

$$\boldsymbol{\pi}_1^{(k)}(1) = m\sum_\nu \int [[(\dot{\mathbf{r}}_\nu^\alpha - \mathbf{u}_\nu^\alpha)(\dot{\mathbf{r}}_\nu^\alpha - \mathbf{u}_\nu^\alpha)]]^\alpha \Psi_\alpha(\mathbf{r}, \mathbf{Q}^\alpha, t)d\mathbf{Q}^\alpha \tag{14.5}$$

$$\boldsymbol{\pi}_2^{(k)}(1) = m\sum_\nu \int (\mathbf{u}_\nu^\alpha - \mathbf{v})(\mathbf{u}_\nu^\alpha - \mathbf{v})\Psi_\alpha(\mathbf{r}, \mathbf{Q}^\alpha, t)d\mathbf{Q}^\alpha \tag{14.6}$$

We now consider these two contributions separately.

When Eq. (12.12) is used for the momentum-space average we obtain: ⟨EMT$_\text{m}$⟩

$$\boldsymbol{\pi}_1^{(k)}(1) = N_\alpha n_\alpha kT\boldsymbol{\delta} \tag{14.7}$$

Note that the term involving the vector $\mathbf{a} = \nabla \ln T$ vanishes.

Next we estimate the magnitude of $\boldsymbol{\pi}_2^{(k)}$; we do this only for an isothermal system at constant composition. For this restricted situation, Eq. (13.14) becomes:

$$\begin{aligned}(\mathbf{u}_\nu^\alpha - \mathbf{v}(\mathbf{r})) &= [\boldsymbol{\kappa}\cdot\Sigma_k B_{\nu k}\mathbf{Q}_k^\alpha] - \frac{1}{\zeta}\Sigma_k \bar{B}_{k\nu}\frac{\partial}{\partial \mathbf{Q}_k^\alpha}(kT\ln\Psi_\alpha + \phi^\alpha)\\ &= [\boldsymbol{\kappa}\cdot\Sigma_j\Sigma_k B_{\nu j}\Omega_{jk}\mathbf{Q}_k'^\alpha] + \frac{H}{\zeta}\Sigma_j\Sigma_k\bar{B}_{j\nu}\Omega_{jk}[(\boldsymbol{\alpha}_k^{-1} - \delta)\cdot\mathbf{Q}_k']\end{aligned} \tag{14.8}$$

in which $\boldsymbol{\alpha}_k$ is given in Eq. (13.10) and $\mathbf{Q}_k'$ in Eq. (13.11); the formula, $\phi^\alpha = \Sigma_k \frac{1}{2} HQ_k^2$, has been used for the intramolecular potential energy. When this expression is substituted into Eq. (14.6), the integrations can be performed and, after considerable manipulation, one obtains: ⟨AFEIRC$_\text{m}$T$_\text{0}$V⟩

$$\begin{aligned}\boldsymbol{\pi}_2^{(k)}(1) = n_\alpha kT(mH/\zeta^2)\Sigma_j a_j[&4\{\lambda_j\boldsymbol{\kappa}\cdot\boldsymbol{\alpha}_j\cdot\lambda_j\boldsymbol{\kappa}^\dagger\}\\ &+ 2\{\boldsymbol{\Gamma}_j\cdot\lambda_j\boldsymbol{\kappa}^\dagger + \lambda_j\boldsymbol{\kappa}\cdot\boldsymbol{\Gamma}_j\} - \{\boldsymbol{\Gamma}_j\cdot(\boldsymbol{\delta} - \boldsymbol{\alpha}_j^{-1})\}]\end{aligned} \tag{14.9}$$

It is thought that, for $\lambda_H\dot{\gamma} < 100$, $\boldsymbol{\pi}_2^{(k)}(1)$ can be neglected with respect to $\boldsymbol{\pi}_1^{(k)}(1)$, since the factor (mH/ζ^2) is probably quite small. For example, according to an estimate of Bird, Fan, and Curtiss [24] for one solution of a polystyrene in an α-chloronaphthalene, the value of (mH/ζ^2) is about $10^{-10}N_\alpha^2$.

Next, we turn to the contribution $\boldsymbol{\pi}^{(\phi)}(1)$ given in Eq. (7.15). For the Rouse chain, the first-order term in the Taylor expansion of this quantity can be written as:

$$\boldsymbol{\pi}^{(\phi)}(1) = -H\Sigma_k \int \mathbf{Q}_k\mathbf{Q}_k\Psi_\alpha d\mathbf{Q}^\alpha = -H\Sigma_k \int \mathbf{Q}'_k\mathbf{Q}'_k\Psi_\alpha d\mathbf{Q}'^\alpha \tag{14.10}$$

When Eq. (13.8) is used for the distribution function and Eq. 13.26 is used to evaluate the integral, we get: ⟨AEFIRXC$_\mathrm{m}$T$_0$V$_\mathrm{m}$⟩

$$\boldsymbol{\pi}^{(\phi)}(1) = -n_\alpha kT\Sigma_k\boldsymbol{\alpha}_k \tag{14.11}$$

Then the final expression for the first-order term in the polymer contribution to the stress tensor is (if we neglect the ostensibly small term in Eq. (14.9)): ⟨AEFIRSXC$_\mathrm{m}$T$_0$V$_\mathrm{m}$⟩

$$\boldsymbol{\pi}_\alpha(1) = N_\alpha n_\alpha kT\boldsymbol{\delta} - n_\alpha kT\Sigma_k\boldsymbol{\alpha}_k \tag{14.12}$$

or

$$\boldsymbol{\tau}_\alpha(1) = \boldsymbol{\pi}_\alpha(1) - \boldsymbol{\pi}_{\alpha,eq} = (N_\alpha - 1)n_\alpha kT\boldsymbol{\delta} - n_\alpha kT\Sigma_k\boldsymbol{\alpha}_k \tag{14.13}$$

This is the same as DPL, Eq. (15.3-17). Thus the stress tensor is given in terms of the Finger strain tensor via Eq. (13.10). The polymer contribution to the stress tensor for the Hookean dumbbell model is obtained by replacing $\Sigma_k\boldsymbol{\alpha}_k$ in Eq. (14.12) by the $\boldsymbol{\alpha}$ tensor defined in Eq. (13.7).

14.1.2 Second-Order Contribution $\boldsymbol{\pi}_\alpha(2)$ to the Stress Tensor

It follows from Eqs. (4.8) and (12.20) that the second-order term for $\boldsymbol{\pi}_1^{(k)}$ is: ⟨EMPT$_\mathrm{m}$⟩

$$\begin{aligned}\boldsymbol{\pi}_1^{(k)}(2) &= -\nabla\cdot kT\Sigma_\nu \int \mathbf{R}_\nu^\alpha(1 + \mathbf{a}\cdot\mathbf{R}_\nu^\alpha)\Psi_\alpha(\mathbf{r},\mathbf{Q}^\alpha,t)d\mathbf{Q}^\alpha\boldsymbol{\delta} \\ &= -(\mathbf{a}\nabla : kT\Sigma_\nu \int \mathbf{R}_\nu^\alpha\mathbf{R}_\nu^\alpha\Psi_\alpha(\mathbf{r},\mathbf{Q}^\alpha,t)d\mathbf{Q}^\alpha)\boldsymbol{\delta}\end{aligned} \tag{14.14}$$

For isothermal systems $\boldsymbol{\pi}_1^{(k)}(2)$ is zero. For nonisothermal solutions of Hookean dumbbells Eq. (13.8) can be used to show that: ⟨DEPT$_\mathrm{m}$⟩

$$\boldsymbol{\pi}_1^{(k)}(2) = -\left(\mathbf{a}\nabla : \frac{n_\alpha(kT)^2}{2H}\boldsymbol{\alpha}\right)\boldsymbol{\delta} \tag{14.15}$$

It follows now from Eq. (13.21) that $\boldsymbol{\pi}_1^{(k)}(2)$ is zero, provided that we neglect terms quadratic in $\mathbf{a}$, $\mathbf{b}_\alpha$, and $\nabla\boldsymbol{\alpha}$. We have not given the expression for $\boldsymbol{\pi}_2^{(k)}(2)$, as it is presumed to be quite small.

The second term of the expansion of Eq. (7.15) gives:

$$\pi^{(\phi)}(2) = -\tfrac{1}{2}\nabla\sum_{\nu}\int \mathbf{R}_{\nu}^{\alpha}\mathbf{R}_{\nu}^{\alpha}\mathbf{F}_{\nu}^{(\phi)\alpha}\Psi_{\alpha}(\mathbf{r},\mathbf{Q}^{\alpha},t)d\mathbf{Q}^{\alpha} \tag{14.16}$$

For isothermal Rouse chains this contribution is zero. For Hookean dumbbells in nonisothermal solutions it follows from Eq. (13.18) that $\boldsymbol{\pi}^{(\phi)}(2)$ is zero.

14.1.3 Third-Order Contribution $\boldsymbol{\pi}_{\alpha}(3)$ to the Stress Tensor

The third-order terms for the contributions to the stress tensor are, for *Hookean dumbbells*:

$$\boldsymbol{\pi}_1^{(k)}(3) = \tfrac{1}{2}\nabla\nabla{:}kT\Sigma_{\nu}\int \mathbf{R}_{\nu}^{\alpha}\mathbf{R}_{\nu}^{\alpha}(1+\mathbf{a}\cdot\mathbf{R}_{\nu}^{\alpha})\boldsymbol{\delta}\Psi_{\alpha}(\mathbf{r},\mathbf{Q}^{\alpha},t)d\mathbf{Q}^{\alpha}$$

$$= \frac{1}{4H}(\nabla\nabla{:}n_{\alpha}(kT)^2\boldsymbol{\alpha})\boldsymbol{\delta} \tag{14.17}$$

$$\boldsymbol{\pi}^{(\phi)}(3) = \tfrac{1}{6}\nabla\nabla{:}\sum_{\nu}\int \mathbf{R}_{\nu}^{\alpha}\mathbf{R}_{\nu}^{\alpha}\mathbf{R}_{\nu}^{\alpha}\mathbf{F}_{\nu}^{(\phi)\alpha}\Psi_{\alpha}(\mathbf{r},\mathbf{Q}^{\alpha},t)d\mathbf{Q}^{\alpha}$$

$$= -\frac{1}{24H}\nabla\nabla{:}n_{\alpha}(kT)^2\mathbf{I}\boldsymbol{\alpha}\mathbf{I} \tag{14.18}$$

in which $\mathbf{I}\boldsymbol{\alpha}\mathbf{I}$ is defined in Eq. 13.32. If terms quadratic in $\mathbf{a}$, $\mathbf{b}_{\alpha}$, and $\nabla\alpha$ are neglected, then $\boldsymbol{\pi}_{\alpha}(3)$ need not be further considered.

14.1.4 The Constitutive Equation for an Isothermal Solution of Rouse Chains

For an isothermal system, by combining Eq. (14.13) with Eq. (13.9) the $\boldsymbol{\alpha}_k$-tensors can be eliminated and a *constitutive equation for the stress tensor* is obtained, through terms of second order: ⟨AEFIRSXC$_m$T$_0$V$_m$⟩

$$\boldsymbol{\tau}_{\alpha} = \Sigma_k\boldsymbol{\tau}_k^{\alpha}; \qquad \boldsymbol{\tau}_k^{\alpha} + \lambda_k\boldsymbol{\tau}_{k(1)}^{\alpha} = -n_{\alpha}kT\lambda_k\dot{\boldsymbol{\gamma}} \tag{14.19}$$

in which $\dot{\boldsymbol{\gamma}} = \nabla\mathbf{v} + (\nabla\mathbf{v})^{\dagger}$ is the rate-of-deformation tensor. From this one can calculate various rheological properties of the model fluid. For example, as shown in Sect. 15.3 of DPL one can calculate the viscosity, which shows no variation with the shear rate $\dot{\gamma}$: ⟨AEFINRSXC$_m$T$_0$V$_m$⟩

$$\eta - \eta_s = n_{\alpha}kT\sum_j\lambda_j = \tfrac{1}{12}n_{\alpha}(N_{\alpha}^2 - 1)(kT\zeta/H) \tag{14.20}$$

This is in conflict with the experimental data, which show a strong shear-rate dependence. Also, this model leads to an infinite elongational viscosity at a finite elongation rate $\dot{\varepsilon}$: ⟨AEFINRSXC$_m$T$_0$V$_m$⟩

$$\bar{\eta} - 3\eta_s = \sum_j \frac{3n_\alpha kT\lambda_j}{(1 + \lambda_j\dot{\varepsilon})(1 - 2\lambda_j\dot{\varepsilon})} \tag{14.21}$$

which is unrealistic. These and other defects mean that these models are not acceptable for use in polymer fluid dynamics. We are using these models here to show how the general formulas can be simplified for specific bead-spring models.

Keep in mind that Eq. (14.12) has been derived by assuming that the velocity gradients are constant over molecular dimensions. However, when we insert the expression for $\boldsymbol{\alpha}_k$ from Sect. 13.3, we must require that the velocity gradients be constant over a longer distance. Nonetheless, it is common practice to use the resulting constitutive equation (e.g. Eq. (14.19)), along with the equations of continuity and motion, to solve flow problems in which the flow is far from homogeneous. Lodge [25] has cautioned against this and has introduced the term "stress calculator" for constitutive equations that are derived for restricted types of flow only. In this connection, see also the discussion following Eq. (13.12) above.

14.2 The Constitutive Equation for an Isothermal Solution of Hookean Dumbbells in a System with Concentration Gradients

We begin by writing Eq. (13.4) for Hookean dumbbells, multiplying this equation by $-H\mathbf{QQ}$, and integrating over $\mathbf{Q}$. Then, when Eqs. (14.10) and (14.16), appropriately simplified for $N_\alpha = 2$ and isothermal systems, are used, we find that – through terms of third order: ⟨ADEFIXT$_0$V$_m$⟩

$$\frac{D}{Dt}\boldsymbol{\pi}^{(\phi)} = \{\boldsymbol{\kappa}\cdot\boldsymbol{\pi}^{(\phi)} + \boldsymbol{\pi}^{(\phi)}\cdot\boldsymbol{\kappa}^\dagger\} + \frac{kT}{2\zeta}\nabla^2\boldsymbol{\pi}^{(\phi)} - \frac{4H}{\zeta}\boldsymbol{\pi}^{(\phi)} - \frac{4n_\alpha kTH}{\zeta}\boldsymbol{\delta} \tag{14.22}$$

This equation can then be rewritten using the definition of the convected derivative defined after Eq. (13.6) as:

$$\boldsymbol{\pi}^{(\phi)} + \lambda_H\boldsymbol{\pi}^{(\phi)}_{\{1\}} = D_{tr}\lambda_H\nabla^2\boldsymbol{\pi}^{(\phi)} - n_\alpha kT\boldsymbol{\delta} \tag{14.23}$$

in which $\lambda_H = \zeta/4H$ is the time constant and $D_{tr} = kT/2\zeta$ is the translational diffusivity (see Eq. (15.7)). Equation (14.23) is the same as Eq. (43) of Bhave, Armstrong, and Brown [26]. Then, using $\boldsymbol{\pi}_\alpha = \boldsymbol{\pi}^{(k)} + \boldsymbol{\pi}^{(\phi)} = 2n_\alpha kT\boldsymbol{\delta} + \boldsymbol{\pi}^{(\phi)}$, we can rewrite Eq. (14.23) as follows: ⟨ADEFISXT$_0$V$_m$⟩

$$\boldsymbol{\pi}_\alpha + \lambda_H\boldsymbol{\pi}_{\alpha(1)} = D_{tr}\lambda_H\nabla^2\boldsymbol{\pi}_\alpha + n_\alpha kT\boldsymbol{\delta} - 2n_\alpha kT\lambda_H\dot{\boldsymbol{\gamma}} + 2kT\boldsymbol{\delta}\left(\frac{Dn_\alpha}{Dt} - D_{tr}\nabla^2 n_\alpha\right) \tag{14.24}$$

When Eq. (13.4), written for Hookean dumbbells, is integrated over **Q**, we obtain: ⟨ADEFIXT$_0$V$_m$⟩

$$\frac{Dn_\alpha}{Dt} = D_{tr}\nabla^2 n_\alpha \tag{14.25}$$

which enables us to drop the last term in Eq. (14.24). Then the latter can be written as: ⟨ADEFISXT$_0$V$_m$⟩

$$\boldsymbol{\tau}_\alpha + \lambda_H \boldsymbol{\tau}_{\alpha(1)} = D_{tr}\lambda_H \nabla^2 \boldsymbol{\tau}_\alpha - n_\alpha k T \lambda_H \dot{\boldsymbol{\gamma}} \tag{14.26}$$

This is the constitutive equation for a solution of Hookean dumbbells in which concentration gradients are present; it includes terms up through third order. The inclusion of a term involving the Laplacian of the stress tensor was first suggested by El-Kareh and Leal [27] to account for diffusion of macromolecules across streamlines. Various other equations containing the Laplacian term have appeared in the literature; these have been compared by Beris and Mavrantzas [30].

14.3 Influence of Temperature Gradients on the Stress Tensor (Hookean Dumbbells)

If the fluid is a system with a temperature gradient, which is constant throughout, then Eq. (14.12) (appropriately simplified for Hookean dumbbells) is still valid as long as Eq. (13.21) is used for the $\boldsymbol{\alpha}$ tensor.

As an illustration, we can consider a dilute solution in which the polymer molecules are modeled as Hookean dumbbells. For a steady shear flow in the x-direction between plates perpendicular to the y-direction, we get from Eq. (13.22) for the shear stress, including terms through the third order: ⟨ADEFMPSXC$_m$T$_c$V⟩

$$\begin{aligned}\pi_{\alpha,yx} = \pi^{(\phi)}_{\alpha,yx} &= -n_\alpha kT(\alpha^{(0)}_{yx} + \alpha^{(1)}_{yx}) \\ &= -n_\alpha kT(\lambda_H\dot{\gamma} - 2a_x y(\lambda_H\dot{\gamma})^2) \\ &= -n_\alpha kT\lambda_H\dot{\gamma}(1 - 2\lambda_H(\mathbf{v}\cdot\mathbf{a}))\end{aligned} \tag{14.27}$$

In the last we have written the result in terms of the flow velocity $\mathbf{v}$ and the quantity $\mathbf{a} = \nabla \ln T$. This result shows that the stresses in the fluid are influenced by the temperature gradient, and that the magnitude of the effect depends on the local fluid velocity. If the imposed temperature gradient is at right angles to the direction of flow, then there is no effect. If the temperature increases in the direction of flow, then the fluid particle located at position r at time t will bring with it a "recollection" of having been in a region of lower temperature during the time period before time t. This means that the shear stress at position $\mathbf{r}$ will be altered because of the memory of the fluid in accordance with the second term in Eq. (13.21). As far as we know, such effects have not been observed or sought for experimentally.

14.4 Charged Rouse Chains in an Electric Field at Equilibrium

We now turn our attention to a dilute solution of charged Rouse chains at equilibrium in a constant electric field of intensity **E**; bead "1" has an electric charge $-q$ and bead "N_α" has a charge $+q$, so that the chain has a dipole moment $\mu_\alpha = q\Sigma_j \mathbf{Q}_j^\alpha$. The intramolecular and external potentials are given by $\phi^\alpha = \frac{1}{2}H\Sigma_j(\mathbf{Q}_j^\alpha \cdot \mathbf{Q}_j^\alpha)$ and $\phi^{(e)\alpha} = -q(\Sigma_j \mathbf{Q}_j^\alpha \cdot \mathbf{E})$. Then the normalized equilibrium distribution function is given by:

$$\Psi_\alpha = n_\alpha \frac{\exp(-(\phi^\alpha + \phi^{(e)\alpha})/kT)}{\int \exp(-(\phi^\alpha + \phi^{(e)\alpha})/kT)d\mathbf{Q}^\alpha} \tag{14.28}$$

The three contributions to the stress tensor are then:

$$\boldsymbol{\pi}^{(k)\alpha} = N_\alpha n_\alpha kT\boldsymbol{\delta} \tag{14.29}$$

$$\boldsymbol{\pi}^{(e)\alpha} = n_\alpha q \int (\Sigma_j \mathbf{Q}_j^\alpha)\mathbf{H}\psi_\alpha d\mathbf{Q}^\alpha \tag{14.30}$$

$$\boldsymbol{\pi}^{(\phi)\alpha} = -n_\alpha H\Sigma_j \int \mathbf{Q}_j^\alpha \mathbf{Q}_j^\alpha \psi_\alpha d\mathbf{Q}^\alpha \tag{14.31}$$

The sum of these last two contributions is:

$$\begin{aligned}\boldsymbol{\pi}^{(\phi)\alpha} + \boldsymbol{\pi}^{(e)\alpha} &= n_\alpha \Sigma_j \int \mathbf{Q}_j^\alpha (q\mathbf{E} - H\mathbf{Q}_j^\alpha)\psi_\alpha d\mathbf{Q}^\alpha \\ &= kT\Sigma_j \int \mathbf{Q}_j^\alpha \frac{\partial \psi_\alpha}{\partial \mathbf{Q}_j^\alpha} d\mathbf{Q}^\alpha \\ &= -kT\Sigma_j \int \psi_\alpha \boldsymbol{\delta} d\mathbf{Q}^\alpha = -kT(N_\alpha - 1)\boldsymbol{\delta}\end{aligned} \tag{14.32}$$

Hence the final result is

$$\boldsymbol{\pi}_\alpha = nkT\boldsymbol{\delta} \tag{14.33}$$

in agreement with Eq. A of Table 13.3-1 of DPL. Therefore for this system one obtains Eq. (14.33) regardless of the intensity of the **E** field. Eq. (14.33) is also obtained for an electrically neutral Rouse chain with charges of $-q$ on all odd-numbered beads and charges of $+q$ on all even numbered beads.

15 The Mass-Flux Vector and the Diffusivity Tensor (DPL, Sect. 18.4)

Equation (6.11) for the mass flux of the diffusing species α can be rewritten as:

$$\mathbf{j}_\alpha(\mathbf{r}, t) = \mathbf{j}_\alpha(1) + \mathbf{j}_\alpha(2) + \mathbf{j}_\alpha(3) + \cdots \tag{15.1}$$

where

$$\mathbf{j}_\alpha(1) = \Sigma_\nu m_\nu^\alpha \int (\mathbf{u}_\nu^\alpha - \mathbf{v})\Psi_\alpha(\mathbf{r}, \mathbf{Q}^\alpha, t)d\mathbf{Q}^\alpha \tag{15.2}$$

$$\mathbf{j}_\alpha(2) = -\nabla\cdot\Sigma_\nu m_\nu^\alpha \int \mathbf{R}_\nu^\alpha(\mathbf{u}_\nu^\alpha - \mathbf{v})\Psi_\alpha(\mathbf{r}, \mathbf{Q}^\alpha, t)d\mathbf{Q}^\alpha \tag{15.3}$$

$$\begin{aligned}\mathbf{j}_\alpha(3) &= \tfrac{1}{2}\nabla\nabla:\Sigma_\nu m_\nu^\alpha \int \mathbf{R}_\nu^\alpha\mathbf{R}_\nu^\alpha(\mathbf{u}_\nu^\alpha - \mathbf{v})\Psi_\alpha(\mathbf{r}, \mathbf{Q}^\alpha, t)d\mathbf{Q}^\alpha \\ &\quad + \tfrac{1}{2}\nabla\nabla:\Sigma_\nu m_\nu^\alpha \int \mathbf{R}_\nu^\alpha\mathbf{R}_\nu^\alpha\mathbf{v}\Psi_\alpha(\mathbf{r}, \mathbf{Q}^\alpha, t)d\mathbf{Q}^\alpha \\ &\quad - \tfrac{1}{2}\mathbf{v}\nabla\nabla:\Sigma_\nu m_\nu^\alpha \int \mathbf{R}_\nu^\alpha\mathbf{R}_\nu^\alpha\Psi_\alpha(\mathbf{r}, \mathbf{Q}^\alpha, t)d\mathbf{Q}^\alpha \\ &= \tfrac{1}{2}\nabla\nabla:\Sigma_\nu m_\nu^\alpha \int \mathbf{R}_\nu^\alpha\mathbf{R}_\nu^\alpha(\mathbf{u}_\nu^\alpha - \mathbf{v})\Psi_\alpha(\mathbf{r}, \mathbf{Q}^\alpha, t)d\mathbf{Q}^\alpha \\ &\quad + [\nabla\cdot\Sigma_\nu m_\nu^\alpha \int \mathbf{R}_\nu^\alpha\mathbf{R}_\nu^\alpha\Psi_\alpha(\mathbf{r}, \mathbf{Q}^\alpha, t)d\mathbf{Q}^\alpha]\cdot\nabla\mathbf{v}\end{aligned} \tag{15.4}$$

The quantity $\mathbf{j}_\alpha(1)$ comes from combining the first and fourth terms in Eq. (6.11). The contribution $\mathbf{j}_\alpha(2)$ comes from the second and fifth terms in Eq. (6.11), recognizing that the $\mathbf{v}$ inserted into Eq. (15.3) does not contribute because of $\Sigma_\nu m_\nu^\alpha \mathbf{R}_\nu^\alpha = 0$. The first two terms in $\mathbf{j}_\alpha(3)$ come from the second term in Eq. (6.11), and the last term in $\mathbf{j}_\alpha(3)$ comes from the sixth term in Eq. (6.11). To obtain the last form of Eq. (15.4) we make use of the fact that for any symmetrical tensor $\mathbf{\Lambda}$

$$\tfrac{1}{2}\nabla\nabla:\mathbf{\Lambda}\mathbf{v} = \tfrac{1}{2}\mathbf{v}\nabla\nabla:\mathbf{\Lambda} + [\nabla\cdot\mathbf{\Lambda}]\cdot\boldsymbol{\kappa}^\dagger + \tfrac{1}{2}\mathbf{\Lambda}:\nabla\boldsymbol{\kappa}^\dagger \tag{15.5}$$

and then we neglect the $\nabla\boldsymbol{\kappa}^\dagger$ term, since the discussion is to be restricted to homogeneous flows. Here $\mathbf{\Lambda}$ stands for the expression following the dot within the brackets in the last line of Eq. 15.4.

We first consider the $\mathbf{j}_\alpha(1)$ term for arbitrary bead-spring models in which all beads have the same mass m and the same friction coefficient ζ; then we investigate the $\mathbf{j}_\alpha(2)$ term for the Rouse chain model. Next we give a derivation of a stress-diffusion relation for the simplified model of Hookean dumbbells (that is, a Rouse chain with $N_\alpha = 2$), which makes use of the $\mathbf{j}_\alpha(2)$ term. Then, we show how the use of the $\mathbf{j}_\alpha(3)$ term leads to a different result. These discussions and Appendix B are helpful in understanding the nature of the series expansions and some of the problems associated with them, because they are not expansions in a physical parameter.

15.1 The Contribution $\mathbf{j}_\alpha(1)$ to the Mass Flux for Arbitrary Bead-Spring Models: Concentration, Thermal, and Forced Diffusion Contributions

We now insert the expression for the average bead velocity in a nonisothermal system, given in Eq. (13.14), into the equation for $\mathbf{j}_\alpha(1)$ above. When the sum on ν is performed and the integrations over the $\mathbf{Q}_k^\alpha$ are carried out, we get finally: $\langle \mathrm{AEFMC_m T_m V_m} \rangle$

$$\mathbf{j}_\alpha(1) = mn_\alpha\left(-\frac{kT}{\zeta}\mathbf{b}_\alpha - \frac{kT}{\zeta}\mathbf{a} + \frac{1}{\zeta}\int \mathbf{F}^{(e)\alpha}\psi_\alpha d\mathbf{Q}^\alpha \right)$$

$$= -\frac{kT}{N_\alpha\zeta}\left(\nabla\rho_\alpha + \rho_\alpha\nabla\ln T - \frac{\rho_\alpha}{kT}\mathbf{G}^\alpha \right) \tag{15.6}$$

in which we have introduced the symbol $\mathbf{G}^\alpha$, closely related to the $\mathbf{G}$ in Eq. (7.11). In going from the first to the second line of Eq. (15.6), we have kept only the first term in the series expression for n_α as a function of ρ_α as discussed at the end of Appendix B.

Equation (15.6) describes, then, the concentration diffusion, the thermal diffusion (the *Soret effect*), and the forced diffusion of polymer species α. The *translational diffusivity* is then given by the relation

$$D_{tr} = \frac{kT}{N_\alpha\zeta} \tag{15.7}$$

This is an unsatisfactory result in that it gives a diffusivity that is inversely proportional to the molecular weight of the polymer. The calculation can be refined by replacing Eq. (12.6) by the expression that includes hydrodynamic interaction, and then one gets Kirkwood's famous result, that the diffusivity is inversely proportional to the square root of the molecular weight. For more details, including literature references and comparisons with experimental data, see DPL, pp. 174–175. For further theoretical developments on diffusion in dilute flowing polymer solutions, see the series of papers by Öttinger [28, 29].

15.2 The Contribution $\mathbf{j}_\alpha(2)$ to the Mass Flux Arising from Velocity Gradients: Rouse Chains in an Isothermal Fluid

We now return to Eq. (15.1) and consider the expression for $\mathbf{j}_\alpha(2)$ specifically for the Rouse chain model in a fluid that is at constant temperature and in which there are no external forces. For this model we can develop the velocity difference appearing in $\mathbf{j}_\alpha(2)$ as follows, by using Eq. (13.14) with $\mathbf{a}$ equal to zero: $\langle \mathrm{AEFRXT_0 V_m} \rangle$

$$[[\dot{\mathbf{r}}_\nu^\alpha - \mathbf{v}]]^\alpha = \left[\boldsymbol{\kappa}\cdot\sum_k B_{\nu k}\mathbf{Q}_k^\alpha\right] - \frac{kT}{N_\alpha\zeta}\nabla\ln n_\alpha$$

$$-\frac{1}{\zeta}\left[\frac{kT}{N_\alpha}\frac{\partial}{\partial \mathbf{r}_c^\alpha}\ln\psi_\alpha + kT\sum_k \bar{B}_{k\nu}\frac{\partial}{\partial \mathbf{Q}_k^\alpha}\ln\psi_\alpha + \sum_k \bar{B}_{k\nu}H\mathbf{Q}_k^\alpha\right]$$

$$= \left[\boldsymbol{\kappa}\cdot\sum_{kl} B_{\nu k}\Omega_{kl}\mathbf{Q}_l'^\alpha\right]$$

$$+\frac{1}{\zeta}\left[-\frac{kT}{N_\alpha}\nabla\ln n_\alpha \mid H\sum_{kl}\bar{B}_{k\nu}\Omega_{kl}[\boldsymbol{\alpha}_l^{-1}-\boldsymbol{\delta}]\cdot\mathbf{Q}_l'^\alpha\right] \quad (15.8)$$

To get the second equality, we made use of Eqs. (13.8), (13.11), and (2.4).

When Eq. (15.8) is substituted into the expression for $\mathbf{j}_\alpha(2)$, the term in Eq. (15.8) containing $\nabla\ln n_\alpha$ does not contribute. The remaining two terms can be developed as follows: $\langle \text{AEFRXT}_0\text{V}\rangle$

$$\mathbf{j}_\alpha(2) = -\nabla\cdot\left[\frac{H}{\zeta}\sum_{\nu ijkl} m_\nu^\alpha B_{\nu i}\bar{B}_{k\nu}\Omega_{ij}\Omega_{kl}\int \mathbf{Q}_j'^\alpha\mathbf{Q}_l'^\alpha\Psi_\alpha d\mathbf{Q}^\alpha\cdot(\boldsymbol{\alpha}_l^{-1}-\boldsymbol{\delta})\right.$$

$$\left. + \sum_{\nu ijkl} m_\nu^\alpha B_{\nu i}B_{\nu k}\Omega_{ij}\Omega_{kl}\int \mathbf{Q}_j'^\alpha\mathbf{Q}_l'^\alpha\Psi_\alpha d\mathbf{Q}^\alpha\cdot\boldsymbol{\kappa}^\dagger\right]$$

$$= -\nabla\cdot m\sum_l \int \mathbf{Q}_l'^\alpha\mathbf{Q}_l'^\alpha\Psi_\alpha d\mathbf{Q}^\alpha\cdot\left[\frac{H}{\zeta}(\boldsymbol{\alpha}_l^{-1}-\boldsymbol{\delta}) + c_l\boldsymbol{\kappa}^\dagger\right]$$

$$= -\nabla\cdot\frac{n_\alpha mkT}{H}\left\{\frac{H}{\zeta}(N_\alpha-1)\boldsymbol{\delta} - \sum_l \boldsymbol{\alpha}_l\cdot\left(\frac{H}{\zeta}\boldsymbol{\delta} - c_l\boldsymbol{\kappa}^\dagger\right)\right\}$$

$$= -\frac{m^\alpha kT}{N_\alpha\zeta}\left[(N_\alpha-1)\boldsymbol{\delta} - \sum_l(\boldsymbol{\delta}-2\lambda_l\boldsymbol{\kappa})\cdot\boldsymbol{\alpha}_l\right]\cdot\nabla n_\alpha \quad (15.9)$$

To get the first expression above we used Eq. (13.11), and to get the second expression we used Eq. 2.10 as well as Eq. 13.12. To evaluate the integrals in the second expression, Eqs. (13.26) to (13.31) were used, and to get the last line we used the fact that the flow is homogeneous and the definition of the time constants given just after Eq. (13.10). The symbol m is used for the mass of a single bead (all beads being identical), and the c_l are the eigenvalues of the Kramers matrix. Then when we use Eqs. (13.10) and (6.7), and also the first term in the series in Eq. (B.21), we finally get for the sum of the first two terms in the Taylor series expansion of the mass flux: $\langle \text{AEFRXT}_0\text{V}\rangle$

$$\mathbf{j}_\alpha = -\left(\frac{kT}{N_\alpha\zeta}\right)\left\{\boldsymbol{\delta} + \sum_l[2\lambda_l\boldsymbol{\kappa} + (\boldsymbol{\delta}-2\lambda_l\boldsymbol{\kappa})\cdot\boldsymbol{\Gamma}_l]\right\}\cdot\nabla\rho_\alpha = -[\boldsymbol{\Delta}_\alpha\cdot\nabla\rho_\alpha] \quad (15.10)$$

Therefore we get for the *diffusivity tensor*: $\langle \text{AEFRXT}_0\text{V}\rangle$

$$\boldsymbol{\Delta}_\alpha = \left(\frac{kT}{N_\alpha\zeta}\right)\left\{\boldsymbol{\delta} + \sum_l[2\lambda_l\boldsymbol{\kappa} + (\boldsymbol{\delta}-2\lambda_l\boldsymbol{\kappa})\cdot\boldsymbol{\Gamma}_l]\right\} \quad (15.11)$$

The second term gives the influence of the flow field on the diffusivity tensor.

We now show how the previous results can lead to a relation between the mass flux and the momentum flux; this is done for Hookean dumbbells only. We begin by specializing the third line of Eq. (15.9) for Hookean dumbbells and make use of Eqs. (14.12), (14.15), and (14.16), written for dumbbells, to get the (nonobjective) expression: ⟨ADEFIXC$_m$T$_0$V⟩

$$\mathbf{j}_\alpha(2) = -\nabla\cdot\left\{\frac{n_\alpha mkT}{H}\left[\frac{H}{\zeta}\boldsymbol{\delta} - \left(\boldsymbol{\delta} - \frac{\boldsymbol{\tau}_\alpha}{n_\alpha kT}\right)\cdot\left(\frac{H}{\zeta}\boldsymbol{\delta} - \frac{1}{2}\nabla\mathbf{v}\right)\right]\right\} \tag{15.12}$$

Here $\boldsymbol{\tau}_\alpha = \boldsymbol{\pi}_\alpha - \boldsymbol{\pi}_{\alpha,eq.} = \boldsymbol{\pi}_\alpha - n_\alpha kT\boldsymbol{\delta}$ is that part of the stress tensor that vanishes when the fluid is at rest; it is understood that this expression for the stress tensor is valid to third order in the Taylor expansion.

Implicit in the derivation of Eq. (15.12) is the assumption of a homogeneous flow field – that is, one in which the velocity gradient (and therefore also the stress tensor and the mass-flux vector) is constant throughout space. If we regard this assumption as valid only over a length scale of the order of the polymer molecules, but allow the flow field to be nonhomogeneous on the scale of the fluid flow pattern, we can proceed to examine Eq. (15.12), on the basis that the stress tensor and velocity gradients may be spatially dependent (see, however, the caveat at the end of Sect. 14.1). If we take that point of view then Eq. (15.12) can be written as: ⟨ADEFIXC$_m$T$_0$V⟩

$$\begin{aligned}\mathbf{j}_\alpha(2) = &-2D_{tr}\lambda_H[\nabla\cdot\rho_\alpha\nabla\mathbf{v}] - (2mD_{tr}/kT)[\nabla\cdot\boldsymbol{\tau}_\alpha] \\ &+ (4mD_{tr}\lambda_H/kT)[\nabla\cdot\{\boldsymbol{\tau}_\alpha\cdot\nabla\mathbf{v}\}]\end{aligned} \tag{15.13}$$

in which $D_{tr} = kT/2\zeta$. This is in agreement with the *stress-diffusion relation* given by Bhave, Armstrong, and Brown [26] in their study of the flow of polymers in nonhomogeneous flow fields. See, however, the more complete relations given in Eq. (15.16) and Eq. (B.17) where the next higher level terms in the Taylor series expansions are retained.

15.3 The Contribution $\mathbf{j}_\alpha(3)$ *to the Mass Flux Arising from Velocity Gradients: Rouse Chains in an Isothermal Fluid*

We substitute the expression for $(\mathbf{u}_\nu^\alpha - \mathbf{v})$ from Eq. (15.8) into $\mathbf{j}_\alpha(3)$, noting that the first and third terms in $(\mathbf{u}_\nu^\alpha - \mathbf{v})$ contribute nothing, in as much as the integrands are odd for the Rouse chain. The remaining term, containing $\nabla \ln n_\alpha$, can be shown to give zero as follows:

$$\tfrac{1}{2}\nabla\nabla{:}\Sigma_\nu m_\nu^\alpha \frac{kT}{N_\alpha\zeta}\left(\int \mathbf{R}_\nu^\alpha\mathbf{R}_\nu^\alpha\Psi_\alpha)(\mathbf{r},Q^\alpha,t)d\mathbf{Q}^\alpha\right)\nabla\ln n_\alpha = \tfrac{1}{2}\nabla\nabla{:}\frac{kT}{H}(\Sigma_k c_k\boldsymbol{\alpha}_k)\mathbf{b}_\alpha \tag{15.14}$$

which is zero for homogeneous flows with constant $\mathbf{b}_\alpha$. Therefore, the final expression for the mass flux, including the contributions from $\mathbf{j}_\alpha(1)$, $\mathbf{j}_\alpha(2)$, and

$\mathbf{j}_\alpha(3)$ is the result in Eq. (15.10) with the velocity-gradient terms deleted. That is,

$$\boldsymbol{\Delta}_\alpha = \left(\frac{kT}{N_\alpha\zeta}\right)\left\{\boldsymbol{\delta} + \sum_k \boldsymbol{\Gamma}_k\right\} = \frac{1}{N_\alpha\zeta n_\alpha}\boldsymbol{\pi}_\alpha \tag{15.15}$$

Here Eqs. (14.12 – 14.18) have been used, which implies that the stress tensor is correct through the third-order terms in the Taylor-series expansion. Note that the diffusion tensor is symmetric, because of the symmetry of the stress tensor. Then for homogeneous flows we finally get the following *stress-diffusion relation* in place of Eq. 15.13: ⟨AEFRXT$_0$Vm⟩

$$\mathbf{j}_\alpha(\mathbf{r},t) = -[\boldsymbol{\Delta}_\alpha\cdot\nabla\rho_\alpha] = \frac{1}{N_\alpha\zeta}\left[\nabla\cdot\frac{\rho_\alpha}{n_\alpha}\boldsymbol{\pi}_\alpha\right] = -\frac{m}{\zeta}[\nabla\cdot\boldsymbol{\pi}_\alpha] \tag{15.16}$$

since, according to Eq. (14.12), the polymer contribution to the stress tensor divided by the number density is independent of position, and therefore can be taken inside the del operator. In the last step in Eq. (15.16) ρ_α has been approximated by $N_\alpha m n_\alpha$ (cf. Eq. (B.21)). This is the (objective) result obtained by Beris and Mavrantzas [30] for Rouse chains by a different sequence of operations. The final result in Eq. (15.16) for Rouse chains turns out to be considerably more general than the derivation implies. In Appendix B it is shown that the same result can be obtained for arbitrary bead-spring models with springs of any kind. See, however, comment iii in Sect. 12.4.

15.4 The Diffusivity Tensor for Steady-state Shear and Elongational Flows

As an illustration of Eq. (15.15) we can consider a steady simple shear flow of the form $v_x = \dot{\gamma}y$, $v_y = 0$, and $v_z = 0$, where $\dot{\gamma}$ is the constant shear rate. For this flow field the diffusion tensor may be obtained by using the $\boldsymbol{\alpha}_j$-tensors for the Rouse model (see end of Sect. 13.3) and the sum of the time constants given in DPL, Eq. 15.3–15.25:

$$\boldsymbol{\Delta}_\alpha = \frac{kT}{N_\alpha\zeta}\begin{bmatrix} 1-\frac{2}{45}(N_\alpha^2-1)(2N_\alpha^2+7)(\zeta\dot{\gamma}/4H)^2 & -\frac{1}{3}(N_\alpha^2-1)(\zeta\dot{\gamma}/4H) & 0 \\ -\frac{1}{3}(N_\alpha^2-1)(\zeta\dot{\gamma}/4H) & 1 & 0 \\ 0 & 0 & 1 \end{bmatrix} \tag{15.17}$$

If the concentration gradient is in the y-direction, then the Rouse model predicts no change in the mass flux in the y-direction, but does give a *cross-diffusion* term in the x-direction. If, on the other hand, the concentration gradient is in the x-direction, then – according to the Rouse chain model – the mass flux in the x-direction is altered by the flow, and there is a cross-diffusion effect in the y-direction.

Similarly for steady elongational flow with $v_x = -\frac{1}{2}\dot{\varepsilon}x$, $v_y = -\frac{1}{2}\dot{\varepsilon}y$, and $v_z = \dot{\varepsilon}z$, we get for very small elongation rates $\dot{\varepsilon}$:

$$\Delta_\alpha = \frac{kT}{N_\alpha \zeta} \times \begin{bmatrix} 1 - \frac{\zeta\dot{\varepsilon}}{12H}(N_\alpha^2 - 1) + \cdots & 0 & 0 \\ 0 & 1 - \frac{\zeta\dot{\varepsilon}}{12H}(N_\alpha^2 - 1) + \cdots & 0 \\ 0 & 0 & 1 + \frac{\zeta\dot{\varepsilon}}{6H}(N_\alpha^2 - 1) - \cdots \end{bmatrix} \tag{15.18}$$

That is, in steady elongational flow the velocity gradients can have an effect on the mass flux.

It should be kept in mind that the Rouse model has many defects, and that therefore the above results may be of limited interest. They are given as an illustration of analytical results that can be obtained from Eq. (6.11). We know of no experimental data with which Eqs. (15.17) and (15.18) can be compared.

16 The Heat-Flux Vector and the Thermal Conductivity (Hookean Dumbbells)

We now apply the results of Sect. 8 to examine the heat flux in a dilute solution in order to find the contribution of the solute species to the thermal conductivity of the fluid. We assume that the contributions of the solvent and solute species are additive; that is, as explained in the introduction to Sect. 14, we do not take into account the polymer-solvent interaction effects, so that $\mathbf{q} = \mathbf{q}_s + \mathbf{q}_\alpha$. We further assume that there are no external forces acting on the polymer molecules.

After simplifying the general expressions for Hookean dumbbells, we proceed to illustrate their use by deriving the heat-flux vector for the Hookean dumbbell model. We conclude with some specific results for shear and elongational flows.

16.1 *Expression for* $(\mathbf{u}_\nu^\alpha - \mathbf{v})$ *in a Nonisothermal System*

We will need the expressions for $\mathbf{q}_1^{(k)}$ (first line in Eq. (8.11)), $\mathbf{q}_2^{(k)}$ (second line in Eq. (8.11)), and $\mathbf{q}^{(\phi)}$ (first line in Eq. (8.18)). We will work out the first, second, and third terms in the Taylor-series expansions of these three expressions. Each of these expressions contains a double-bracket factor containing $(\dot{\mathbf{r}}_\nu^\alpha - \mathbf{v})$ in which $\mathbf{v}$ is the fluid velocity at $\mathbf{r}$. We replace this quantity by $(\dot{\mathbf{r}}_\nu^\alpha - \mathbf{u}_\nu^\alpha) + (\mathbf{u}_\nu^\alpha - \mathbf{v})$; the first of these terms depends on the velocities (or momenta) and coordinates of the beads, and the second depends solely on the coordinates.

An expression for $(\mathbf{u}_\nu^\alpha - \mathbf{v})$ appropriate for multicomponent, nonisothermal systems was given in Eq. (13.14). For any elastic dumbbell this quantity can be written as follows for $\nu = 1, 2$: $\langle \mathrm{ADEFXC_m T_m V_m} \rangle$

$$\mathbf{u}_\nu^\alpha(\mathbf{r}, \mathbf{Q}, t) - \mathbf{v}(\mathbf{r}, t) = \left[-\frac{1}{\zeta} \bar{B}_{1\nu} \left(\frac{\partial}{\partial \mathbf{Q}} \phi^\alpha + kT \frac{\partial}{\partial \mathbf{Q}} \ln \psi_\alpha \right) + \boldsymbol{\kappa} \cdot B_{\nu 1} \mathbf{Q} \right]$$
$$- \frac{kT}{\zeta} \left[\left(\mathbf{a} + \frac{1}{2} \mathbf{b}_\alpha \right) + \mathbf{a} \cdot \bar{B}_{1\nu} B_{\nu 1} \mathbf{Q} \frac{\partial}{\partial \mathbf{Q}} \ln \psi_\alpha + \frac{1}{2} \frac{\partial}{\partial \mathbf{r}_c} \ln \psi_\alpha \right] \tag{16.1}$$

For Hookean dumbbells the derivatives of $\ln \psi_\alpha$ are found from Eq. (13.18):

$$\frac{\partial}{\partial \mathbf{Q}} \ln \psi_\alpha = -\frac{H}{kT} [\boldsymbol{\alpha}^{-1} \cdot \mathbf{Q}] \tag{16.2}$$

$$\frac{\partial}{\partial \mathbf{r}_c} \ln \psi_\alpha = \left[\frac{1}{2} \nabla \ln \det \left(\frac{H}{2kT} \boldsymbol{\alpha}^{-1} \right) - \left(\nabla \frac{H}{2kT} \boldsymbol{\alpha}^{-1} \right) : \mathbf{QQ} \right]$$
$$= \left[\frac{1}{2} \left(\nabla \frac{H}{2kT} \boldsymbol{\alpha}^{-1} \right) : \left(\frac{2kT}{H} \boldsymbol{\alpha} \right) - \left(\nabla \frac{H}{2kT} \boldsymbol{\alpha}^{-1} \right) : \mathbf{QQ} \right]$$
$$= \frac{1}{2} (\nabla \boldsymbol{\alpha}^{-1} - \mathbf{a} \boldsymbol{\alpha}^{-1}) : \left\{ \boldsymbol{\alpha} - \frac{H}{kT} \mathbf{QQ} \right\}$$
$$= -\frac{1}{2} \left[(\nabla \boldsymbol{\alpha}) : \left\{ \boldsymbol{\alpha}^{-1} - \frac{H}{kT} \boldsymbol{\alpha}^{-1} \cdot \mathbf{QQ} \cdot \boldsymbol{\alpha}^{-1} \right\} \right.$$
$$\left. + \left[3 - \frac{H}{kT} (\boldsymbol{\alpha}^{-1} : \mathbf{QQ}) \right] \mathbf{a} \right] \tag{16.3}$$

From the last three equations, and also Eqs. (2.6) and (2.7), we get

$$\mathbf{u}_\nu^\alpha(\mathbf{r}, \mathbf{Q}^\alpha, t) - \mathbf{v}(\mathbf{r}, t) = (-1)^{\nu - 1} (H/\zeta) [(\boldsymbol{\delta} - \boldsymbol{\alpha}^{-1} - 2\lambda_H \boldsymbol{\kappa}) \cdot \mathbf{Q}]$$
$$- \frac{kT}{\zeta} \left[\left(\frac{1}{4} \mathbf{a} + \frac{1}{2} \mathbf{b}_\alpha \right) - \frac{H}{2kT} [\boldsymbol{\alpha}^{-1} \cdot \mathbf{QQ} \cdot \mathbf{a}] + \frac{H}{4kT} (\boldsymbol{\alpha}^{-1} : \mathbf{QQ}) \mathbf{a} \right.$$
$$\left. - \frac{1}{4} (\nabla \boldsymbol{\alpha}) : \boldsymbol{\alpha}^{-1} + \frac{H}{4kT} (\nabla \boldsymbol{\alpha}) : [\boldsymbol{\alpha}^{-1} \cdot \mathbf{QQ} \cdot \boldsymbol{\alpha}^{-1}\} \right]$$
$$\equiv (-1)^{\nu - 1} (H/\zeta) [\boldsymbol{\Omega} \cdot \mathbf{Q}] - (kT/\zeta)(\mathbf{Y}_1 + \mathbf{Y}_2 + \mathbf{Y}_3 + \mathbf{Y}_4 + \mathbf{Y}_5) \tag{16.4}$$

in which $\mathbf{Y}_1$, $\mathbf{Y}_2$, etc. are the five vectors in the bracket multiplying (kT/ζ) in the second and third lines, and $\boldsymbol{\Omega} = \boldsymbol{\delta} - \boldsymbol{\alpha}^{-1} - 2\lambda_H \boldsymbol{\kappa}$. For systems without velocity gradients $\boldsymbol{\alpha} = \boldsymbol{\delta}$ and $\boldsymbol{\kappa} = 0$, so that $\mathbf{Y}_2$ and $\mathbf{Y}_3$ become considerably simpler, and $\mathbf{Y}_4$, $\mathbf{Y}_5$ and Ω vanish.

16.2 The Taylor-Series Expansion of $\mathbf{q}_1^{(k)}$ *and its Relation to* $\mathbf{j}_\alpha$

The double-bracket factor in the $\mathbf{q}_1^{(k)}$ contribution to the heat flux, given in the first line of Eq. (8.11), has already been evaluated in Eq. (12.21). Therefore, $\mathbf{q}_1^{(k)}$ can be written as follows, provided that we neglect the last term in Eq. (12.21):

$$\mathbf{q}_1^{(k)} = \tfrac{1}{2}m \sum_{\nu=1}^{2} \int 5kT(\mathbf{r} - \mathbf{R}_\nu^\alpha, t)(1 + \mathbf{a}\cdot\mathbf{R}_\nu^\alpha)(\mathbf{u}_\nu^\alpha - \mathbf{v})\Psi_\alpha(\mathbf{r} - \mathbf{R}_\nu^\alpha, \mathbf{Q}, t)d\mathbf{Q} \quad (16.5)$$

As pointed out in connection with Eq. (5.9), the double-bracket quantities are always functions of the same variables as the associated configuration-space distribution function. It is for this reason that the T in Eq. (16.5) must be written as a function of the displaced coordinate $(\mathbf{r} - \mathbf{R}_\nu^\alpha)$. However, according to Eq. (12.3), $T(\mathbf{r} - \mathbf{R}_\nu^\alpha, t)$ can be replaced by $T(\mathbf{r}, t)(1 - \mathbf{a}\cdot\mathbf{R}_\nu^\alpha)$ and then $T(\mathbf{r}, t)$ can be removed from the integral. In addition, terms quadratic in $\mathbf{a}$ are neglected. Then Eq. (16.5) becomes:

$$\mathbf{q}_1^{(k)} = \tfrac{5}{2}kT \sum_{\nu=1}^{2} \int (\mathbf{u}_\nu^\alpha - \mathbf{v})\Psi_\alpha(\mathbf{r} - \mathbf{R}_\nu^\alpha, \mathbf{Q}, t)d\mathbf{Q} \quad (16.6)$$

Comparison of this equation with Eq. (6.10) shows that there is a simple relation between mass flux and the flux of kinetic energy by diffusion:

$$\mathbf{q}_1^{(k)} = \frac{5kT}{2m}\mathbf{j}_\alpha \quad (16.7)$$

That is, the contribution $\mathbf{q}_1^{(k)}$ to the part of the heat-flux vector associated with species α is proportional to the mass flux of species α. Therefore, the Taylor-series results given in Eqs. (15.2–15.4) for the mass-flux vector can be taken over here. When Eq. (16.4) is used in these expressions, we get for the first three terms in the Taylor series for $\mathbf{q}_1^{(k)}$: $\langle \mathrm{ADEFPXC_mT_mV_m} \rangle$

$$\mathbf{q}_1^{(k)}(1) = -5\frac{n_\alpha(kT)^2}{\zeta}\int(\Sigma_j \mathbf{Y}_j)\psi_\alpha d\mathbf{Q} \quad (16.8)$$

$$\mathbf{q}_1^{(k)}(2) = \tfrac{5}{2}kT\nabla\cdot\frac{n_\alpha kT}{\zeta}\int\frac{H}{kT}\{\mathbf{QQ}\cdot\boldsymbol{\Omega}^\dagger\}\psi_\alpha d\mathbf{Q} \quad (16.9)$$

$$\mathbf{q}_1^{(k)}(3) = \tfrac{5}{4}kT[\nabla\cdot n_\alpha\int\mathbf{QQ}\psi_\alpha d\mathbf{Q}]\cdot\nabla\mathbf{v} \quad (16.10)$$

In obtaining these expressions we have used $-\mathbf{R}_1^\alpha = +\mathbf{R}_2^\alpha = \tfrac{1}{2}\mathbf{Q}$. This completes the development of the expressions for $\mathbf{q}_1^{(k)}$ for Hookean dumbbells.

16.3 The Taylor-Series Expansion of $\mathbf{q}_2^{(k)}$

Next we write out the first three terms in the Taylor-series expansion of $\mathbf{q}_2^{(k)}$ starting with the second line of Eq. (8.11): $\langle \mathrm{ADEFPXC_mT_mV_m} \rangle$

$$\mathbf{q}_2^{(k)}(1) = \tfrac{1}{2}n_\alpha \Sigma_\nu \Sigma_\mu \int \phi^\alpha_{\nu\mu}(\mathbf{u}^\alpha_\nu - \mathbf{v})\psi_\alpha d\mathbf{Q} \tag{16.11}$$

$$\mathbf{q}_2^{(k)}(2) = -\tfrac{1}{2}[\nabla\cdot n_\alpha \Sigma_\nu \Sigma_\mu \int \mathbf{R}^\alpha_\nu \phi^\alpha_{\nu\mu}(\mathbf{u}^\alpha_\nu - \mathbf{v})\psi_\alpha d\mathbf{Q}]$$

$$-\tfrac{1}{2}[n_\alpha \Sigma_\nu \Sigma_\mu \int \mathbf{R}^\alpha_\nu \phi^\alpha_{\nu\mu}\psi_\alpha\, d\mathbf{Q}\cdot\nabla\mathbf{v}] \tag{16.12}$$

$$\mathbf{q}_2^{(k)}(3) = \tfrac{1}{4}[\nabla\nabla{:}n_\alpha \Sigma_\nu \Sigma_\mu \int \mathbf{R}^\alpha_\nu \mathbf{R}^\alpha_\nu \phi^\alpha_{\nu\mu}(\mathbf{u}^\alpha_\nu - \mathbf{v})\psi_\alpha d\mathbf{Q}]$$

$$+\tfrac{1}{2}[\nabla\cdot n_\alpha \Sigma_\nu \Sigma_\mu \int \mathbf{R}^\alpha_\nu \mathbf{R}^\alpha_\nu \phi^\alpha_{\nu\mu}\psi_\alpha\, d\mathbf{Q}]\cdot\nabla\mathbf{v} \tag{16.13}$$

Then, when Eq. (16.4) is used for $(\mathbf{u}^\alpha_\nu - \mathbf{v})$ we get:

$$\mathbf{q}_2^{(k)}(1) = -\frac{n_\alpha(kT)^2}{\zeta}\frac{H}{2kT}\int Q^2(\Sigma_j \mathbf{Y}_j)\psi_\alpha d\mathbf{Q} \tag{16.14}$$

$$\mathbf{q}_2^{(k)}(2) = \nabla\cdot\left\{\frac{n_\alpha(kT)^2}{\zeta}\left(\frac{H}{2kT}\right)^2\int Q^2\mathbf{Q}\mathbf{Q}\psi_\alpha d\mathbf{Q}\cdot\mathbf{\Omega}^\dagger\right\} \tag{16.15}$$

$$\mathbf{q}_2^{(k)}(3) = -\tfrac{1}{8}\nabla\nabla{:}\frac{n_\alpha(kT)^2}{\zeta}\left(\frac{H}{2kT}\right)\int Q^2\mathbf{Q}\mathbf{Q}\psi_\alpha d\mathbf{Q}$$

$$+\tfrac{1}{4}\left[\nabla\cdot n_\alpha kT\left(\frac{H}{2kT}\right)\int Q^2\mathbf{Q}\mathbf{Q}\psi_\alpha d\mathbf{Q}\right]\cdot\nabla\mathbf{v} \tag{16.16}$$

This concludes the tabulation of the terms in the Taylor-series for $\mathbf{q}_2^{(k)}$.

16.4 The Taylor-Series Expansion of $\mathbf{q}^{(\phi)}$

Finally, we write out the first three terms in the Taylor expansion for the contribution $\mathbf{q}^{(\phi)}$: $\langle \mathrm{ADEFPXC_mT_mV_m} \rangle$

$$\mathbf{q}^{(\phi)}(1) = \tfrac{1}{2}n_\alpha \Sigma_\nu \Sigma_\mu \int [\mathbf{R}^\alpha_\nu \mathbf{F}^{(\phi)\alpha}_{\nu\mu}\cdot[(\mathbf{u}^\alpha_\nu - \mathbf{v}) + (\mathbf{u}^\alpha_\mu - \mathbf{v})]\psi_\alpha d\mathbf{Q} \tag{16.17}$$

$$\mathbf{q}^{(\phi)}(2) = -\tfrac{1}{4}\cdot n_\alpha \Sigma_\nu \Sigma_\mu \int [\mathbf{R}^\alpha_\nu \mathbf{R}^\alpha_\nu \mathbf{F}^{(\phi)\alpha}_{\nu\mu}\cdot[(\mathbf{u}^\alpha_\nu - \mathbf{v}) + (\mathbf{u}^\alpha_\mu - \mathbf{v})]\psi_\alpha d\mathbf{Q}$$

$$-\tfrac{1}{2}(n_\alpha \Sigma_\nu \Sigma_\mu \int \mathbf{R}^\alpha_\nu \mathbf{R}^\alpha_\nu \mathbf{F}^{(\phi)\alpha}_{\nu\mu}\psi_\alpha d\mathbf{Q})\cdot(\nabla\mathbf{v})^\dagger \tag{16.18}$$

$$\mathbf{q}^{(\phi)}(3) = \tfrac{1}{12}\nabla\nabla{:}n_\alpha \Sigma_\nu \Sigma_\mu \int [\mathbf{R}^\alpha_\nu \mathbf{R}^\alpha_\nu \mathbf{R}^\alpha_\nu \mathbf{F}^{(\phi)\alpha}_{\nu\mu}\cdot[(\mathbf{u}^\alpha_\nu - \mathbf{v}) + (\mathbf{u}^\alpha_\mu - \mathbf{v})]\psi_\alpha d\mathbf{Q}$$

$$+\tfrac{1}{3}(\nabla\cdot n_\alpha \Sigma_\nu \Sigma_\mu \int \mathbf{R}^\alpha_\nu \mathbf{R}^\alpha_\nu \mathbf{R}^\alpha_\nu \mathbf{F}^{(\phi)\alpha}_{\nu\mu}\psi_\alpha d\mathbf{Q}){:}(\nabla\mathbf{v})^\dagger \tag{16.19}$$

Then, using Eq. (16.4) leads to:

$$\mathbf{q}^{(\phi)}(1) = \frac{n_\alpha (kT)^2}{\zeta}\left(\frac{H}{kT}\right)\int [\mathbf{QQ}\cdot(\Sigma_j \mathbf{Y}_j)]\,\psi_\alpha d\mathbf{Q} \tag{16.20}$$

$$\mathbf{q}^{(\phi)}(2) = 0 \tag{16.21}$$

$$\mathbf{q}^{(\phi)}(3) = \tfrac{1}{12}\nabla\nabla{:}\frac{n_\alpha (kT)^2}{\zeta}\left(\frac{H}{kT}\right)\int [\mathbf{QQQQ}\cdot(\Sigma_j \mathbf{Y}_j)]\,\psi_\alpha d\mathbf{Q}$$
$$-\tfrac{1}{6}(\nabla\cdot n_\alpha (kT)\left(\frac{H}{kT}\right)\int \mathbf{QQQQ}\,\psi_\alpha d\mathbf{Q}){:}(\nabla\mathbf{v})^\dagger \tag{16.22}$$

In the above we have used $\mathbf{F}_{12}^{(\phi)\alpha} = -\mathbf{F}_{21}^{(\phi)\alpha} = H\mathbf{Q}$ and $-\mathbf{R}_1^\alpha = +\mathbf{R}_2^\alpha = \frac{1}{2}\mathbf{Q}$.

16.5 *The Contributions* $\mathbf{q}_\alpha(1)$, $\mathbf{q}_\alpha(2)$, *and* $\mathbf{q}_\alpha(3)$ *to the Heat-Flux Vector*

Substitution of the quantities $\mathbf{Y}_j$ and $\mathbf{\Omega}$, defined in Eq. (16.4), into Eqs. (16.8–16.10), (16.14–16.16), and (16.20–16.22) then give the contributions to the heat-flux vector. It then remains to perform the indicated integrations, using the distribution function given in Eq. (13.18). The integrals needed are given in Eqs. (13.26–13.31). After some tedious manipulations, we get: ⟨ADEFPXCTV⟩

$$\mathbf{q}_\alpha(1) = -\frac{n_\alpha (kT)^2}{\zeta}[(\tfrac{5}{2}\boldsymbol{\delta} + \tfrac{1}{2}(tr\boldsymbol{\alpha})\boldsymbol{\delta} + \tfrac{1}{2}\boldsymbol{\alpha})\cdot\mathbf{a} + \nabla\cdot(\tfrac{1}{4}(tr\boldsymbol{\alpha})\boldsymbol{\delta} - \tfrac{1}{2}\boldsymbol{\alpha})$$
$$+ (\tfrac{5}{2}\boldsymbol{\delta} + \tfrac{1}{4}(tr\boldsymbol{\alpha})\boldsymbol{\delta} - \tfrac{1}{2}\boldsymbol{\alpha})\cdot\mathbf{b}_\alpha] \tag{16.23}$$

$$\mathbf{q}_\alpha(2) = +\tfrac{5}{2}kT\left[\nabla\cdot\frac{n_\alpha kT}{\zeta}\{\boldsymbol{\alpha}\cdot\mathbf{\Omega}^\dagger\}\right]$$
$$+\left[\nabla\cdot\frac{n_\alpha (kT)^2}{\zeta}\{(\tfrac{1}{4}(tr\boldsymbol{\alpha})\boldsymbol{\delta} + \tfrac{1}{2}\boldsymbol{\alpha})\cdot\boldsymbol{\alpha}\cdot\mathbf{\Omega}^\dagger\}\right] \tag{16.24}$$

$$\mathbf{q}_\alpha(3) = +5kT\left[\nabla\cdot\frac{n_\alpha kT}{\zeta}\{\boldsymbol{\alpha}\cdot\lambda_H\boldsymbol{\kappa}^\dagger\}\right]$$
$$+\left[\nabla\cdot\frac{n_\alpha (kT)^2}{\zeta}\{(\tfrac{1}{2}(tr\boldsymbol{\alpha})\boldsymbol{\delta} + \boldsymbol{\alpha})\cdot\boldsymbol{\alpha}\cdot\lambda_H\boldsymbol{\kappa}^\dagger\}\right]$$
$$-\tfrac{1}{3}\left[\nabla\cdot\frac{n_\alpha (kT)^2}{\zeta}\{\tfrac{1}{2}\boldsymbol{\alpha}\boldsymbol{\alpha}{:}\lambda_H\dot{\boldsymbol{\gamma}} + \boldsymbol{\alpha}\cdot\lambda_H\dot{\boldsymbol{\gamma}}\cdot\boldsymbol{\alpha}\}\right] \tag{16.25}$$

The first two terms in Eq. (16.25), which contain $\lambda_H\boldsymbol{\kappa}^\dagger$, exactly cancel parts of the two terms in Eq. (16.24), since $\mathbf{\Omega}^\dagger = \boldsymbol{\alpha} - \boldsymbol{\delta} - 2\lambda_H\boldsymbol{\kappa}^\dagger$. Fourth- and higher-order terms in the Taylor expansion are not needed, as long as we neglect spatial derivatives of the temperature, concentration, and velocity gradients, as well as products of the temperature and concentration gradients.

Next, one performs the ∇-operations in Eqs. (16.24) and (16.25), thereby generating terms proportional to $\mathbf{a}$ and $\mathbf{b}_\alpha$. When the contributions from the first, second, and third order are added, we get finally for the contribution to the heat flux from the solute molecules, modeled as Hookean dumbbells: ⟨ADEFPXCTV⟩

$$\mathbf{q}_\alpha = \frac{n_\alpha(kT)^2}{\zeta}[(5\boldsymbol{\delta} + 4\boldsymbol{\alpha} - 2\boldsymbol{\Phi})\cdot\mathbf{a} - \nabla\cdot\boldsymbol{\Phi} + (5\boldsymbol{\delta} - \boldsymbol{\Phi})\cdot\mathbf{b}_\alpha] \tag{16.26}$$

in which

$$\boldsymbol{\Phi} = \tfrac{5}{2}\boldsymbol{\alpha} - \tfrac{1}{2}(tr\boldsymbol{\alpha})\boldsymbol{\delta} + \tfrac{1}{4}(tr\boldsymbol{\alpha})\boldsymbol{\alpha} + \tfrac{1}{2}\{\boldsymbol{\alpha}\cdot\boldsymbol{\alpha}\} - \tfrac{1}{6}\{\boldsymbol{\alpha}\boldsymbol{\alpha}:\lambda_H\dot{\boldsymbol{\gamma}}\} - \tfrac{1}{3}\{\boldsymbol{\alpha}\cdot\lambda_H\dot{\boldsymbol{\gamma}}\cdot\boldsymbol{\alpha}\} \tag{16.27}$$

and $\boldsymbol{\alpha}$ is the quantity defined in Eq. (13.21); see also Eqs. (13.22) to (13.25) for explicit expressions for $\boldsymbol{\alpha}$ for steady-state shear and elongational flows. One can obtain a stress-thermal relation by eliminating $\boldsymbol{\alpha}$ in favor of the stress tensor by using Eq. (14.12); however the relation is far from simple.

The first two terms in Eq. (16.26) give the heat flux associated with a temperature gradient, and the last term is the heat flux associated with a concentration gradient. The velocity-gradient dependence is contained in the tensors $\dot{\boldsymbol{\gamma}}$ and $\boldsymbol{\alpha} = \boldsymbol{\alpha}^{(0)} + \boldsymbol{\alpha}^{(1)}$ (see Eqs. 13.20–13.25). Note that in the ∇-terms it is $\boldsymbol{\alpha}^{(1)}$ that contributes (since it is porportional to $\mathbf{a}$), whereas elsewhere in the equation only $\boldsymbol{\alpha}^{(0)}$ need be considered (in as much as it is linear in $\mathbf{a}$ and $\mathbf{b}_\alpha$).

16.6 The Heat-Flux Vector in Nonflow Systems

When $\nabla\mathbf{v} = \mathbf{0}$ (so that $\boldsymbol{\alpha} = \boldsymbol{\delta}$ and $\boldsymbol{\Omega} = \mathbf{0}$), the contributions in Eqs. (16.9), (16.10), (16.15), (16.16), and (16.22) are zero. Then Eqs. (16.8), (16.9), and (16.20) lead to: ⟨ADEFPXCTV$_0$⟩

$$\mathbf{q}_1^{(k)} = -n_\alpha(k^2T^2/\zeta)(\tfrac{5}{2}\mathbf{a} + \tfrac{5}{2}\mathbf{b}_\alpha) \tag{16.28}$$

$$\mathbf{q}_2^{(k)} = -n_\alpha(k^2T^2/\zeta)(\mathbf{a} + \tfrac{3}{4}\mathbf{b}_\alpha) \tag{16.29}$$

$$\mathbf{q}^{(\phi)} = -n_\alpha(k^2T^2/\zeta)(\mathbf{a} - \tfrac{1}{2}\mathbf{b}_\alpha) \tag{16.30}$$

When added together these give for the polymer contribution to the heat-flux vector: ⟨ADEFPXCTV$_0$⟩

$$\mathbf{q}_\alpha = -n_\alpha(k^2T^2/\zeta)(\tfrac{9}{2}\mathbf{a} + \tfrac{11}{4}\mathbf{b}_\alpha) \tag{16.31}$$

If we now eliminate $\mathbf{b}_\alpha$ in this expression by using Eq. (15.6) (or Eq. (C.26)), we get:

$$\mathbf{q}_\alpha = -\frac{7}{4}\frac{n_\alpha k^2T^2}{\zeta}\mathbf{a} + \frac{11}{2}\frac{kT}{m_m^\alpha}\mathbf{j}_\alpha \tag{16.32}$$

The contribution of the solute to the *thermal conductivity* of the solution is then defined as the coefficient of the temperature gradient in Eq. (16.32). If we neglect polymer–solvent interaction, then the thermal conductivity becomes: ⟨ADEFNPXCTV$_0$⟩

$$\lambda = \lambda_s + \tfrac{7}{4} n_\alpha (k^2 T/\zeta) \tag{16.33}$$

This definition of thermal conductivity is consistent with that given in Refs. [30a], [11], and [12]. It should be noted that the spring constant H does not appear in the result.

Van den Brule [31, Dissertation, Eq. (a.39), p. 76] was the first to obtain an expression for the thermal conductivity for Hookean dumbbells. His result, for small mH/ζ^2, has the same form as Eq. (16.33), but with 7/4 replaced by 3/2. However, in his theory he did not take into account all three contributions to the heat-flux vector. Results analogous to Eqs. (16, 28–16, 33) have also been obtained for FENE dumbbells [31a].

16.7 The Heat-flux Vector in Steady-state Shear and Elongational Flows

For any kind of flow we may write the polymer contribution to the heat-flux vector as:

$$\mathbf{q}^\alpha = -[\boldsymbol{\beta}_\alpha \cdot \nabla T] - [\boldsymbol{\sigma}_\alpha \cdot \nabla n_\alpha] \tag{16.34}$$

The tensorial coefficient $\boldsymbol{\beta}_\alpha$ in this expression is *not* the "thermal conductvity tensor," since it is not a generalization of the thermal conductivity (scalar) in Eq. (16.33). The two tensors appearing in Eq. (16.34) are, however, a convenient representation for the fluid response in a system with velocity gradients as well as temperature and concentration gradients.

For *steady shear flow* with $v_x = \dot{\gamma} y$, $v_y = 0$, and $v_z = 0$, these tensors can be displayed explicitly by using Eqs. (13.22) and (13.23) to give us directly:

$$\beta_\alpha = \frac{n_\alpha k^2 T}{\zeta} \begin{bmatrix} \frac{9}{2} + \frac{31}{12}\lambda_H^2 \dot{\gamma}^2 + \frac{1}{6}\lambda_H^4 \dot{\gamma}^4 & -\frac{23}{6}\lambda_H \dot{\gamma} - \frac{1}{3}\lambda_H^3 \dot{\gamma}^3 & 0 \\ -\frac{1}{3}\lambda_H \dot{\gamma} - \frac{7}{3}\lambda_H^3 \dot{\gamma}^3 & \frac{9}{2} + 2\lambda_H^2 \dot{\gamma}^2 & 0 \\ 0 & 0 & \frac{9}{2} + \frac{5}{3}\lambda_H^2 \dot{\gamma}^2 \end{bmatrix} \tag{16.35}$$

$$\sigma_\alpha = \frac{k^2 T^2}{\zeta} \begin{bmatrix} \frac{11}{4} - \frac{15}{2}\lambda_H^2 \dot{\gamma}^2 - \lambda_H^4 \dot{\gamma}^4 & -\frac{47}{12}\lambda_H \dot{\gamma} - \frac{1}{6}\lambda_H^3 \dot{\gamma}^3 & 0 \\ -\frac{47}{12}\lambda_H \dot{\gamma} - \frac{1}{6}\lambda_H^3 \dot{\gamma}^3 & \frac{11}{4} + \lambda_H^2 \dot{\gamma}^2 & 0 \\ 0 & 0 & \frac{11}{4} + \frac{5}{6}\lambda_H^2 \dot{\gamma}^2 \end{bmatrix} \tag{16.36}$$

Several comments can be made about these results:

a. If a temperature gradient in the x-direction is imposed, heat will flow in both the x- and y-directions. The component $\beta_{\alpha,xx}$ is even in $\lambda_H\dot{\gamma}$, in accordance with the notion that the heat flow in the x-direction should be independent of the direction of shearing. On the other hand, the component $\beta_{\alpha,yx}$ is odd, which means that the heat flow in the y-direction will reverse its direction when the direction of shearing is reversed.

b. If a temperature gradient is imposed in the y-direction, comments similar to those in (a) regarding the reversal of the direction of flow can be made.

c. Note that for small shear rates the heat flow in the x-direction in (a) is greater than the heat flow in the y-direction in (b), for the temperature gradients of the same magnitude. In other words, with increasing shear rate, the heat flow in the direction of polymer stretching is greater than that perpendicular to the stretching direction.

For *steady elongational flow* with $v_x = -\frac{1}{2}\dot{\varepsilon}x$, $v_y = -\frac{1}{2}\dot{\varepsilon}y$, and $v_z = \dot{\varepsilon}z$ the two tensors defined in Eq. 16.33 can be displayed explicitly by making use of Eqs. (13.24) and (13.25). Both tensors are diagonal, with $\boldsymbol{\beta}_\alpha$ being given by

$$\begin{aligned}\beta_{\alpha,xx} = \beta_{\alpha,yy} = \frac{n_\alpha k^2 T}{\zeta}\Bigg(5 &+ \frac{1}{1+\lambda_H\dot{\varepsilon}} + \frac{1}{1-2\lambda_H\dot{\varepsilon}} - \frac{2(1+\frac{2}{3}\lambda_H\dot{\varepsilon})}{(1+\lambda_H\dot{\varepsilon})^2} \\ &- \frac{\frac{1}{2}(1-\frac{4}{5}\lambda_H\dot{\varepsilon})}{(1+\lambda_H\dot{\varepsilon})(1-2\lambda_H\dot{\varepsilon})} - \frac{\frac{3}{4}\lambda_H\dot{\varepsilon}}{(1+\lambda_H\dot{\varepsilon})(1+\frac{1}{2}\lambda_H\dot{\varepsilon})} \\ &+ \frac{\frac{1}{4}\lambda_H\dot{\varepsilon}}{(1-2\lambda_H\dot{\varepsilon})(1-\frac{5}{2}\lambda_H\dot{\varepsilon})} - \frac{\lambda_H\dot{\varepsilon}(1+\frac{2}{3}\lambda_H\dot{\varepsilon})}{(1+\lambda_H\dot{\varepsilon})^2(1+\frac{1}{2}\lambda_H\dot{\varepsilon})} \\ &- \frac{\frac{1}{8}\lambda_H\dot{\varepsilon}(1-\frac{4}{3}\lambda_H\dot{\varepsilon})}{(1+\lambda_H\dot{\varepsilon})(1-2\lambda_H\dot{\varepsilon})(1-\frac{5}{2}\lambda_H\dot{\varepsilon})} \\ &- \frac{\frac{1}{8}\lambda_H\dot{\varepsilon}(1-\frac{4}{3}\lambda_H\dot{\varepsilon})}{(1+\lambda_H\dot{\varepsilon})(1-2\lambda_H\dot{\varepsilon})(1+\frac{1}{2}\lambda_H\dot{\varepsilon})}\Bigg)\end{aligned} \tag{16.37}$$

$$\begin{aligned}\beta_{\alpha,zz} = \frac{n_\alpha k^2 T}{\zeta}\Bigg(5 &+ \frac{2}{1+\lambda_H\dot{\varepsilon}} - \frac{\frac{3}{2}(1-\frac{4}{3}\lambda_H\dot{\varepsilon})}{(1-2\lambda_H\dot{\varepsilon})^2} - \frac{1-\frac{2}{3}\lambda_H\dot{\varepsilon}}{(1+\lambda_H\dot{\varepsilon})(1-2\lambda_H\dot{\varepsilon})} \\ &+ \frac{2\lambda_H\dot{\varepsilon}}{(1-2\lambda_H\dot{\varepsilon})(1-\lambda_H\dot{\varepsilon})} - \frac{\lambda_H\dot{\varepsilon}}{(1+\lambda_H\dot{\varepsilon})(1+2\lambda_H\dot{\varepsilon})} \\ &+ \frac{\frac{3}{2}\lambda_H\dot{\varepsilon}(1-\frac{4}{3}\lambda_H\dot{\varepsilon})}{(1-2\lambda_H\dot{\varepsilon})^2(1-\lambda_H\dot{\varepsilon})} + \frac{\frac{1}{2}\lambda_H\dot{\varepsilon}(1+\frac{2}{3}\lambda_H\dot{\varepsilon})}{(1+\lambda_H\dot{\varepsilon})(1-2\lambda_H\dot{\varepsilon})(1+2\lambda_H\dot{\varepsilon})} \\ &+ \frac{\frac{1}{2}\lambda_H\dot{\varepsilon}(1+\frac{2}{3}\lambda_H\dot{\varepsilon})}{(1+\lambda_H\dot{\varepsilon})(1-2\lambda_H\dot{\varepsilon})(1-\lambda_H\dot{\varepsilon})}\Bigg)\end{aligned} \tag{16.38}$$

and σ_α by

$$\sigma_{\alpha,xx} = \sigma_{\alpha,yy} = \frac{(kT)^2}{\zeta}\left(\frac{\frac{11}{4} + \frac{47}{2}\lambda_H\dot{\varepsilon} - \frac{59}{6}\lambda_H^2\dot{\varepsilon}^2 - 10\lambda_H^3\dot{\varepsilon}^3}{(1 - 2\lambda_H\dot{\varepsilon})(1 + \lambda_H\dot{\varepsilon})^2}\right) \tag{16.39}$$

$$\sigma_{\alpha,zz} = \frac{(kT)^2}{\zeta}\left(\frac{\frac{11}{4} - \frac{193}{12}\lambda_H\dot{\varepsilon} + \frac{29}{3}\lambda_H^2\dot{\varepsilon}^2 + 20\lambda_H^3\dot{\varepsilon}^3}{(1 - 2\lambda_H\dot{\varepsilon})^2(1 + \lambda_H\dot{\varepsilon})}\right) \tag{16.40}$$

Furthermore, for small $\lambda_H\dot{\varepsilon}$ the components of $\boldsymbol{\beta}_\alpha$ are:

$$\beta_{\alpha,xx} = \frac{n_\alpha k^2 T}{\zeta}\left(\tfrac{9}{2} + \tfrac{25}{12}\lambda_H\dot{\varepsilon} + \cdots\right) \tag{16.41}$$

$$\beta_{\alpha,zz} = \frac{n_\alpha k^2 T}{\zeta}\left(\tfrac{9}{2} - \tfrac{25}{6}\lambda_H\dot{\varepsilon} + \cdots\right) \tag{16.42}$$

At low elongational rates the heat flux in the direction of stretching decreases, whereas that in the transverse direction increases. For larger values of the elongation rate both components increase, ultimately becoming infinite. This unphysical behavior is presumably associated with the unrealistic infinite stretching of the dumbbells.

For oriented polymer *solids* enhanced thermal conductivity in the direction of orientation has been measured by Issi and collaborators [32, 33].

17 Concluding Comments and Suggestions for Further Research

In Table 1 we have summarized the contributions to the flux expressions as given by the first term in the Taylor-series expansion of each of the fluxes; this tabulation emphasizes some of the similarities among the various fluxes. In Table 2 we give more general expressions, from which higher-order terms in the Taylor series can be generated. These expressions are needed primarily for the study of the cross-effects.

The discussion up through Sect. 12 involves only two major assumptions: the use of bead-spring models (i.e., models with no internal constraints) and the assumption of pairwise additive potentials between the beads in the system. In order to go further with the theory, five major additional assumptions were introduced, assumptions that have been tacitly made in almost all kinetic theory work so far:

a. The "short-range force assumption" given in Eq. (12.1) means that $\mathbf{q}^{(d)}$ and $\boldsymbol{\pi}^{(d)}$ need not be further considered.

b. The assumption in Eq. (12.9) of a Maxwellian velocity distribution about $\mathbf{u}_\nu^\alpha$ eliminates the need for knowing the phase-space distribution function f_α. This assumption has also been called "the assumption of equilibration in momentum space."

c. The use of a simple linear relation between hydrodynamic force and relative velocity in Eq. (12.6) obviates the need for evaluating the integral in Eq. (11.6), which involves the pair distribution function. Alternative empiricisms involving hydrodynamic interaction, nonisotropic friction coefficients, and Basset forces have not been explored here.

d. The "omission of the acceleration term" in the equation of motion for the beads, (Eq. 11.4), enables us to use the simple force balance given in Eq. (12.5).

e. Linear gradients in velocity, temperature, and concentration have be assumed, as in Eqs. (12.2), (12.3), and (12.4).

These five major assumptions can and should be challenged. It is hoped that Brownian dynamics or molecular dynamics calculations can give us the means for assessing the seriousness of these assumptions or replacing them with less restrictive assumptions.

In order to make use of the flux expressions in Sects. 6, 7, and 8, it is necessary to have the singlet distribution function and – unless the short-range force assumption is used – the doublet distribution function as well. Virtually nothing is known about the doublet distribution function. If we knew how to make a reasonable guess of this function (possibly obtainable from molecular or Brownian dynamics), then we could estimate the contributions to the fluxes in Table 1 that involve the molecule-molecule interactions.

Even for the singlet distribution function, as pointed out in Sect. 13, we have very few solutions of Eq. (13.4). Only for the Rouse linear chain model and the Hookean dumbbell model have complete analytical solutions been obtained. More needs to be done for models involving nonlinear springs or for models that are not chains (e.g., stars, rings, combs, and branched polymers) [33a, 33b, 33c]. Some major advance in the solution of many-dimensional partial differential equations will have to occur before any progress on this front is forthcoming. Here again, Brownian or molecular dynamics simulations will surely give useful information before analytical progress is made. With these simulations the rigorous expressions for the mass-flux vector, the stress tensor, and the heat-flux vector can be used, and progress along this line would appear to be promising.

The illustrative applications in Sects. 14–16 to the momenum, mass, and energy fluxes are given for the Hookean dumbbell and Rouse models. These models are known to be very inadequate because of their infinite extensibility. They have been used solely to show how to apply the general formulas of Tables 1 and 2. The next phase of study should emphasize calculations using bead-spring chain models with finitely extensible springs, using molecular or Brownian dynamics [23, 33d, 33e, 33f].

There are also major experimental challenges in this field. Systematic studies of the effects of velocity fields on the diffusivity and thermal conductivity tensors would be most welcome. Regrettably there are virtually no experimental data on the various "cross effects" for which theoretical formulas are now available. Perhaps this review will stimulate some laboratory studies of these various phenomena.

Relatively little use has been made of the phase-space kinetic theory to study solvent effects in polymer solution dynamics. Also much more can be done with regard to wall effects, flow of polymers in constrictions, behavior of polymers at interfaces, and the thermal and diffusional properties of polyelectrolytes.

Much more needs to be done in the area of optical and electrical properties. Dielectric relaxation, light scattering, and birefringence need to be treated analogously to the phase-space discussion given here for the transport properties.

Acknowledgements. The authors are very much indebted to Prof. H.C. Öttinger (Eidgenössische Technische Hochschule Zürich) for many stimulating discussions and incisive criticisms; we have profited very much from his interest in our manuscript as it progressed through several stages. In addition we wish to acknowledge the following persons for interesting correspondence and suggestions for improving the manuscript: Prof. A.N. Beris (University of Delaware), Dr. G.D.C. Kuiken (Technische Universiteit Delft), Prof. J.D. Schieber (Illinois Institute of Technology), Prof. Emer. A.S. Lodge (University of Wisconsin), Dr. B.H.A.A. van den Brule (Shell-Rijswijk and Technische Universiteit Delft), Dr. R.J.J. Jongschaap (Universiteit Twente), Prof. R.C. Armstrong (MIT), Prof. J.M. Wiest (University of Alabama), Prof. M.W. Johnson, Jr. (University of Wisconsin), and Prof. J.J. de Pablo (University of Wisconsin). One of the authors (RBB) wishes particularly to thank the faculty and staff of the J.M. Burgerscentrum of the Technische Universiteit Delft for their hospitality during the spring semester of 1994 when an earlier version of this manuscript was being developed in connection with a course on the fluid dynamics and kinetic theory of polymeric liquids. RBB particularly enjoyed interacting with the university and industry participants in this course.

Appendix A
The Stress Tensor at Equilibrium

In Sect. 7 we obtained from nonequilibrium statistical mechanics an expression for the various contributions to the stress tensor $\boldsymbol{\pi}$ (Eqs. (7.8), (7.12), (7.15), and (7.19)). Two questions arise regarding these results:

(i) One could clearly add to each of the contributions a divergenceless tensor; hence the uniqueness of our results can be called into question. Adding a divergenceless tensor would not affect the equation of motion, but would give different results for the force on a surface element, which is $[\mathbf{n}\cdot\boldsymbol{\pi}]\,dS$.

(ii) One could identify the $\mathbf{S}^{(e)}$ in (Eq. 7.9) with the $\mathbf{G}$ in Eq. (7.1) and set $\boldsymbol{\pi}^{(e)}$ equal to zero in Eq. (7.12). This would not change any solution to the equation of motion Eq. (7.1), but it would lead to a different expression for the force, $[\mathbf{n}\cdot\boldsymbol{\pi}]\,dS$ on a surface element.

To investigate these points, we derive an expression for the stress tensor at equilibrium by applying the tensor virial theorem [34] to the polymeric liquid, modeled as interacting bead-spring systems. The discussion here is restricted to systems for which the external forces on the beads are independent of position, and the total external force on a single molecule is zero (an example would be elastic dumbbells with a positive charge on one bead and an equal negative charge on the other in a constant electric field, and with negligible gravitational forces). It is shown that the stress tensor thus obtained is identical to that in Sect. 7, and that neither of the changes in (i) and (ii) can be permitted at equilibrium.

We begin by noting that for any bead in the system the time derivative of the dyad $\mathbf{r}_\nu^{\alpha i}\mathbf{p}_\nu^{\alpha i}$ along a trajectory in the phase space is:

$$\frac{d}{dt}\mathbf{r}_\nu^{\alpha i}\mathbf{p}_\nu^{\alpha i} = \frac{1}{m_\nu^\alpha}\mathbf{p}_\nu^{\alpha i}\mathbf{p}_\nu^{\alpha i} + \mathbf{r}_\nu^{\alpha i}\tilde{\mathbf{F}}_\nu^{\alpha i} = \frac{1}{m_\nu^\alpha}\mathbf{p}_\nu^{\alpha i}\mathbf{p}_\nu^{\alpha i} + \mathbf{r}_\nu^{\alpha i}(\mathbf{F}_\nu^{\alpha i} + \mathbf{F}_\nu^{(w)\alpha i}) \tag{A.1}$$

Here $\tilde{\mathbf{F}}_\nu^{\alpha i}$ is the total force on a bead, including $\mathbf{F}_\nu^{\alpha i}$ (the sum of the intramolecular, intermolecular, and external forces) the force $\mathbf{F}_\nu^{(w)\alpha i}$ due to the walls of the container enclosing the liquid.

We now sum Eq. (A.1) over all beads in the system and integrate over time from 0 to τ:

$$(1/\tau)\sum_{\alpha i\nu}\mathbf{r}_\nu^{\alpha i}\mathbf{p}_\nu^{\alpha i}\big|_0^\tau = \mathbf{K} + \boldsymbol{\Xi} + \boldsymbol{\Xi}^{(w)} \tag{A.2}$$

The term on the left is the difference between the values at time τ and time 0. The terms on the right are:

$$\mathbf{K} = \sum_{\alpha i\nu}\frac{1}{m_\nu^\alpha}\langle\mathbf{p}_\nu^{\alpha i}\mathbf{p}_\nu^{\alpha i}\rangle_\tau \tag{A.3}$$

$$\boldsymbol{\Xi} = \sum_{\alpha i\nu}\langle\mathbf{r}_\nu^{\alpha i}\mathbf{F}_\nu^{\alpha i}\rangle_\tau \tag{A.4}$$

$$\boldsymbol{\Xi}^{(w)} = \sum_{\alpha i\nu}\langle\mathbf{r}_\nu^{\alpha i}\mathbf{F}_\nu^{(w)\alpha i}\rangle_\tau \tag{A.5}$$

in which $\langle\cdots\rangle_\tau$ indicates a time average over the time period from 0 to τ.

If the system is at equilibrium, the term on the left becomes zero in the limit that $\tau\to\infty$, because the position vectors are bounded by the limits of the container, and the momenta will be extremely large only over very small time intervals. In this same limit, the time averages of the first two terms on the right side may be replaced by the phase-space averages, using the equilibrium distribution function. The first term (from (Eq. A.3)) is:

$$\mathbf{K} = \sum_{\alpha i\nu}\frac{1}{m_\nu^\alpha}\langle\mathbf{p}_\nu^{\alpha i}\mathbf{p}_\nu^{\alpha i}\rangle = \sum_{\alpha i\nu}\frac{1}{m_\nu^\alpha}\int\mathbf{p}_\nu^{\alpha i}\mathbf{p}_\nu^{\alpha i}f_{eq}(x)\,dx \tag{A.6}$$

The second term (from (Eq. A.4)) is a tensor form of the "virial of the forces" multiplied by 2 (see [12], pp. 42 and 134]:

$$\begin{aligned}\boldsymbol{\Xi} &= \sum_{\alpha i\nu}\langle\mathbf{r}_\nu^{\alpha i}\mathbf{F}_\nu^{\alpha i}\rangle = \sum_{\alpha i\nu}\int\mathbf{r}_\nu^{\alpha i}\,\mathbf{F}_\nu^{\alpha i}f_{eq}(x)\,dx\\ &= \sum_{\alpha i\nu}\int\mathbf{r}_\nu^{\alpha i}\mathbf{F}_\nu^{\alpha i}\Psi_{eq}(r)\,dr\end{aligned} \tag{A.7}$$

where, once again, the average is with respect to the equilibrium distribution function. In these equations x is the complete set of phase-space variables and r is the complete set of configuration-space variables, as defined in Sect. 3.

To evaluate the third term (in Eq. (A.5), we note that the time average of the sum of all the forces on the beads due to the wall is just the negative of the time-average force on the wall by the beads. From continuum arguments, this force (per unit area) is $[\mathbf{n}\cdot\boldsymbol{\pi}]$, where $\mathbf{n}$ is the unit normal to the wall and π is the stress tensor at the wall.

It was pointed out at the beginning of this appendix that the expression for the stress tensor introduced in Sect. 7 is ambiguous for two reasons. First, one could add a divergenceless tensor to the expression, and second, one could add (or subtract) any multiple of $\boldsymbol{\pi}^{(e)}$, as defined in Eq. (7.12), provided that one changes the external force term $\mathbf{G}$ in Eq. (7.11) appropriately. Neither change would affect the equation of motion, but either of them would affect the expression for the force on a wall, as discussed above. To show that the choice of the expression for $\boldsymbol{\pi}$ discussed in Sect. 7 is the correct choice, we now define a quantity $\hat{\boldsymbol{\pi}}$ as follows:

$$\hat{\boldsymbol{\pi}} = \boldsymbol{\pi} + c_1\boldsymbol{\pi}^{(e)} + c_2\boldsymbol{\beta} \tag{A.8}$$

in which $\boldsymbol{\pi}$ is the sum of four contributions as discussed in Sect. 7 (see immediately after Eq. 7.4), $\boldsymbol{\pi}^{(e)}$ is given by Eq. (7.12), $\boldsymbol{\beta}$ is a divergenceless tensor, and c_1 and c_2 are arbitrary constants. If $\hat{\boldsymbol{\pi}}$ is taken to be the stress tensor, then the force per unit area on a wall is $[\mathbf{n}\cdot\hat{\boldsymbol{\pi}}]$, and thus:

$$\boldsymbol{\Xi}^{(w)} = -\int \mathbf{r}[\mathbf{n}\cdot\hat{\boldsymbol{\pi}}]\,dS = -\left\{\int \mathbf{n}\cdot\hat{\boldsymbol{\pi}}\mathbf{r}\,dS\right\}^{\dagger} \tag{A.9}$$

were dS is a surface element on the walls of the container. At equilibrium $\boldsymbol{\pi}$ and $\boldsymbol{\pi}^{(e)}$ are constant throughout the system, except possibly very near the walls, where the expressions for these tensors vary with the position $\mathbf{r}$, and where the configurations of the molecules are restricted and there may be short-range wall potentials. Because of the smoothness of these changes in $\hat{\boldsymbol{\pi}}$ near the wall, Gauss's theorem may be used to transform the surface integral into a volume integral over the contents of the container:

$$\boldsymbol{\Xi}^{(w)} = -\left\{\int \nabla\cdot\hat{\boldsymbol{\pi}}\mathbf{r}\,d\mathbf{r}\right\}^{\dagger} = -\int \mathbf{r}[\nabla\cdot\hat{\boldsymbol{\pi}}]\,d\mathbf{r} - \int \hat{\boldsymbol{\pi}}\,d\mathbf{r} \tag{A.10}$$

in which $d\mathbf{r}$ is an element of volume in the three-dimensional space. Next we note that for the system under consideration – a fluid at rest with the total external force per molecule being zero – the equation of motion in Eq. (7.1) simplifies to $[\nabla\cdot\boldsymbol{\pi}] = \mathbf{0}$. Substitution into Eq. (A.10) for $\hat{\boldsymbol{\pi}}$ from Eq. (A.8) and using the simplified equation of motion then gives:

$$\boldsymbol{\Xi}^{(w)} = -c_1\int \mathbf{r}[\nabla\cdot\boldsymbol{\pi}^{(e)}]\,d\mathbf{r} - \int(\boldsymbol{\pi} + c_1\boldsymbol{\pi}^{(e)} + c_2\boldsymbol{\beta})\,d\mathbf{r} \tag{A.11}$$

When Eq. (A.2) is used (with the left side equal to zero), Eq. (A.11) becomes:

$$-(\mathbf{K}+\boldsymbol{\Xi}) = -c_1\int \mathbf{r}[\nabla\cdot\boldsymbol{\pi}^{(e)}]\,d\mathbf{r} - \int \boldsymbol{\pi}d\mathbf{r} - c_1\int \boldsymbol{\pi}^{(e)}d\mathbf{r} - c_2\int \boldsymbol{\beta}d\mathbf{r} \tag{A.12}$$

We show presently that the left side of the equation is exactly equal to $-\int \boldsymbol{\pi}d\mathbf{r}$, and therefore Eq. (A.12) becomes:

$$\mathbf{0} = -c_1\int \mathbf{r}[\nabla\cdot\boldsymbol{\pi}^{(e)}]\,d\mathbf{r} - c_1\boldsymbol{\pi}_b^{(e)}V - c_1\int(\boldsymbol{\pi}^{(e)}-\boldsymbol{\pi}_b^{(e)})\,d\mathbf{r} - c_2\int \boldsymbol{\beta}d\mathbf{r} \tag{A.13}$$

in which the subscript b indicates the value in the bulk of the fluid (that is, not in the thin layer near the walls).

From Eq. (A.13) we want to determine c_1 and c_2. The second term on the right depends on the magnitude of V but not on the shape of the container. The values of the first, third, and fourth terms, on the other hand, do depend on the shape of the container, because of the factor $\mathbf{r}$ in the first integral, and because the integrands deviate from the bulk values near the walls in the integral terms. We now distinguish between two cases:

(a) Case I: $\boldsymbol{\beta}$ is a constant tensor

In this case the fourth term on the right becomes $-c_2\boldsymbol{\beta}V$ and is independent of the shape of the container. Therefore the sum of the second and fourth terms may be set equal to zero (the shape-independent terms), and the sum of the first and third terms may be set equal to zero (the shape-dependent terms). It can be concluded from the latter relation that c_1 is zero. Then, from the first relation, it can be concluded that c_2 is zero.

(b) Case II: $\boldsymbol{\beta}$ is not a constant tensor

In this case the second term is the only shape-independent term, and setting it equal to zero gives c_1 equal to zero. Then setting the sum of the shape-dependent terms equal to zero gives c_2 equal to zero.

We are thus led to the conclusion that $\hat{\boldsymbol{\pi}}$ and $\boldsymbol{\pi}$ are the same.

It remains to be shown that the left side of Eq. (A.12) is exactly equal to $-\int \boldsymbol{\pi}d\mathbf{r}$, that is, that:

$$\int \boldsymbol{\pi}d\mathbf{r} = \sum_{\alpha i\nu}\frac{1}{m_\nu^\alpha}\int \mathbf{p}_\nu^{\alpha i}\mathbf{p}_\nu^{\alpha i}f_{eq}(x)\,dx + \sum_{\alpha i\nu}\int \mathbf{r}_\nu^{\alpha i}\mathbf{F}_\nu^{\alpha i}\Psi_{eq}(r)\,dr \tag{A.14}$$

Because we have restricted the discussion to systems for which the total external force on a molecule is zero, we can replace $\mathbf{r}_\nu^{\alpha i}$ by $\mathbf{R}_\nu^{\alpha i}$ in the last term:

$$\int \boldsymbol{\pi}d\mathbf{r} = \sum_{\alpha i\nu}\frac{1}{m_\nu^\alpha}\int \mathbf{p}_\nu^{\alpha i}\mathbf{p}_\nu^{\alpha i}f_{eq}(x)\,dx + \sum_{\alpha i\nu}\int \mathbf{R}_\nu^{\alpha i}\mathbf{F}_\nu^{\alpha i}\Psi_{eq}(r)\,dr \tag{A.15}$$

We now have to show that the right hand side of Eq. (A.15) is the same as the integral (over the volume) of the stress tensor expression derived in Sect. 7, when the latter is evaluated at equilibrium.

The first term on the right side of Eq. (A.15) may be rewritten by introducing the integral over a Dirac delta function:

$$\sum_{\alpha i \nu} \frac{1}{m_\nu^\alpha} \int \mathbf{p}_\nu^{\alpha i} \mathbf{p}_\nu^{\alpha i} f_{eq}(x)\, dx = \sum_{\alpha i \nu} \frac{1}{m_\nu^\alpha} \iint \mathbf{p}_\nu^{\alpha i} \mathbf{p}_\nu^{\alpha i} f_{eq}(x)\, \delta(\mathbf{r}_\nu^{\alpha i} - \mathbf{r}) d\mathbf{r} dx \tag{A.16}$$

When the order of integration is interchanged the right hand side of this equation is exactly equal to the volume integral of the first term on the right side of Eq. (7.5) (in which $\mathbf{v}$ has been set equal to zero). That is, it is equal to the volume integral of $\boldsymbol{\pi}^{(k)}$ for a stationary system at equilibrium.

The second term on the right side of Eq. (A.15) has to be written as the sum of three terms, one for intramolecular forces, one for intermolecular forces, and one for external forces. We go through the details here for the intramolecular force contribution. We begin by introducing the integral over a Dirac delta function:

$$\begin{aligned}
&\sum_{\alpha i \nu} \int \mathbf{R}_\nu^{\alpha i} \mathbf{F}_\nu^{(\phi)\alpha i} \Psi_{eq}(r)\, dr \\
&= \sum_{\alpha i \nu} \iint \mathbf{R}_\nu^{\alpha i} \mathbf{F}_\nu^{(\phi)\alpha i} \Psi_{eq}(r)\, \delta(\mathbf{r}^{\alpha i} - \mathbf{r}^\alpha) d\mathbf{r}^\alpha dr \\
&= \sum_{\alpha \nu} \int \mathbf{R}_\nu^{\alpha} \mathbf{F}_\nu^{(\phi)\alpha} \bar{\Psi}_{\alpha, eq}(\mathbf{r}^\alpha)\, d\mathbf{r}^\alpha \qquad \text{Used Eq. (4.3)} \\
&= \sum_{\alpha \nu} \iint \mathbf{R}_\nu^{\alpha} \mathbf{F}_\nu^{(\phi)\alpha} \Psi_{\alpha, eq}(\mathbf{Q}^\alpha)\, d\mathbf{r}_c^\alpha d\mathbf{Q}^\alpha \qquad \text{Used Eq. (4.4)}
\end{aligned} \tag{A.17}$$

When the order of integration is interchanged, this is the same as the volume integral of the expression for $\boldsymbol{\pi}^{(\phi)}$ given in Eq. (7.15), when the latter is written for a stationary system at equilibrium.

Exactly the same set of steps may be carried out for the external force contribution, and this then leads to a volume integral over the expression for $\boldsymbol{\pi}^{(e)}$ in Eq. (7.12) for a system at equilibrium. A similar development can be made for the intermolecular force contribution; in the first step the integrals over two delta functions are inserted, and then use is made of Eqs. (4.7) and (4.13). This leads to the volume integral of the expression for $\boldsymbol{\pi}^{(d)}$ in Eq. (7.19) for a system at equilibrium.

The final conclusion is that the expression for the stress tensor developed in Sect. 7 is consistent with the tensor virial theorem. Furthermore, it is not permitted to add a divergenceless term to the stress tensor, and the partitioning of the external source term in Eq. (7.9) into an external force term $\mathbf{G}$ and the divergence of $\boldsymbol{\pi}^{(e)}$ in Eq. (7.10) is correct.

Appendix B
A Stress–Diffusion Relation for General Bead–Spring Models

Here we address the question of whether it is possible to get a relation between the mass-flux vector and the polymer contribution to the stress tensor for any bead-spring model in which all beads have mass m and friction coefficient ζ. We consider systems in which intermolecular forces can be neglected.

By applying Eq. (5.7) to the expressions for ρ_α (Eq. (6.6)) and $\rho_\alpha \mathbf{v}_\alpha$ (implied by Eqs. (6.1) and (6.8)), we get: ⟨M⟩

$$\rho_\alpha = m\Sigma_\nu \int \bar{\Psi}_\alpha(\mathbf{r}^\alpha_\nu, t)\,\delta(\mathbf{r}^\alpha_\nu - \mathbf{r})\,d\mathbf{r}^\alpha \tag{B.1}$$

$$\rho_\alpha \mathbf{v}_\alpha = \Sigma_\nu \int [[\mathbf{p}^\alpha_\nu]]^\alpha \bar{\Psi}_\alpha(\mathbf{r}^\alpha_\nu, t)\,\delta(\mathbf{r}^\alpha_\nu - \mathbf{r})\,d\mathbf{r}^\alpha \tag{B.2}$$

so that: ⟨M⟩

$$\begin{aligned}\mathbf{j}_\alpha &= \Sigma_\nu \int [[\mathbf{p}^\alpha_\nu - m\mathbf{v}(\mathbf{r})]]^\alpha \bar{\Psi}_\alpha(\mathbf{r}^\alpha_\nu, t)\,\delta(\mathbf{r}^\alpha_\nu - \mathbf{r})\,d\mathbf{r}^\alpha \\ &= m\Sigma_\nu \int [[(\mathbf{p}^\alpha_\nu/m) - \mathbf{v}(\mathbf{r}^\alpha_\nu)]]^\alpha \bar{\Psi}_\alpha(\mathbf{r}^\alpha_\nu, t)\delta(\mathbf{r}^\alpha_\nu - \mathbf{r})\,d\mathbf{r}^\alpha\end{aligned} \tag{B.3}$$

Because of the Dirac delta function, the argument of $\mathbf{v}$ can be changed from $\mathbf{r}$ to $\mathbf{r}^\alpha_\nu$.

Next, we rewrite the "momentum source" term of Eq. (7.4) (omitting $\mathbf{S}^{(d)}$) by using the force balance of Eq. (12.5); we do this for species α only and apply Eq. (5.7): ⟨AS⟩

$$\begin{aligned}\mathbf{S}^\alpha &= \Sigma_\nu \int (\mathbf{F}^{(\phi)\alpha}_\nu + \mathbf{F}^{(e)\alpha}_\nu)\,\bar{\Psi}_\alpha(\mathbf{r}^\alpha_\nu, t)\,\delta(\mathbf{r}^\alpha_\nu - \mathbf{r})\,d\mathbf{r}^\alpha \\ &= -\Sigma_\nu \int (\mathbf{F}^{(h)\alpha}_\nu + \mathbf{F}^{(b)\alpha}_\nu)\bar{\Psi}_\alpha(\mathbf{r}^\alpha_\nu, t)\,\delta(\mathbf{r}^\alpha_\nu - \mathbf{r})\,d\mathbf{r}^\alpha \\ &\equiv \mathbf{S}^{(h)\alpha} + \mathbf{S}^{(b)\alpha}\end{aligned} \tag{B.4}$$

The hydrodynamic force contribution $\mathbf{S}^{(h)\alpha}$ is given by: ⟨FM⟩

$$\mathbf{S}^{(h)\alpha} = \zeta\Sigma_\nu \int [\mathbf{u}^\alpha_\nu - \mathbf{v}(\mathbf{r}^\alpha_\nu)]\,\bar{\Psi}_\alpha(\mathbf{r}^\alpha_\nu, t)\,\delta(\mathbf{r}^\alpha_\nu - \mathbf{r})\,d\mathbf{r}^\alpha \tag{B.5}$$

where Eq. (12.6) has been used. Comparison of Eqs. (B.3), (B.5), and (B.4) reveals that: ⟨AFMS⟩

$$\mathbf{j}_\alpha = \frac{m}{\zeta}\mathbf{S}^{(h)\alpha} = \frac{m}{\zeta}(\mathbf{S}^\alpha - \mathbf{S}^{(b)\alpha}) \tag{B.6}$$

Because of the simplicity of this relation, it is possible to obtain a relation between the mass-flux vector and the polymer contribution to the stress tensor.

First we note that, by using Eqs. (5.9) and (5.11), the two parts of the first line of Eq. (B.4) can be written as

$$\mathbf{S}^{(\phi)\alpha} = \Sigma_\nu \int \mathbf{F}_\nu^{(\phi)\alpha} \Psi_\alpha(\mathbf{r} - \mathbf{R}_\nu^\alpha, \mathbf{Q}^\alpha, t) d\mathbf{Q}^\alpha = -[\nabla \cdot \boldsymbol{\pi}^{(\phi)\alpha}] \tag{B.7}$$

$$\mathbf{S}^{(e)\alpha} = \Sigma_\nu \int \mathbf{F}_\nu^{(e)\alpha} \Psi_\alpha(\mathbf{r} - \mathbf{R}_\nu^\alpha, \mathbf{Q}^\alpha, t) d\mathbf{Q}^\alpha = \mathbf{G}^\alpha - [\nabla \cdot \boldsymbol{\pi}^{(e)\alpha}] \tag{B.8}$$

where

$$\boldsymbol{\pi}^{(\phi)\alpha} = \iint_0^1 \mathbf{R}_\nu^\alpha \mathbf{F}_\nu^{(\phi)\alpha} \Psi_\alpha(\mathbf{r} - \xi\mathbf{R}_\nu^\alpha, \mathbf{Q}^\alpha, t)\, d\xi d\mathbf{Q}^\alpha \tag{B.9}$$

$$\boldsymbol{\pi}^{(e)\alpha} = \iint_0^1 \mathbf{R}_\nu^\alpha \mathbf{F}_\nu^{(e)\alpha} \Psi_\alpha(\mathbf{r} - \xi\mathbf{R}_\nu^\alpha, \mathbf{Q}^\alpha, t)\, d\xi d\mathbf{Q}^\alpha \tag{B.10}$$

$$\mathbf{G}^\alpha = \int \mathbf{F}^{(e)\alpha} \Psi_\alpha(\mathbf{r}, \mathbf{Q}^\alpha, t)\, d\mathbf{Q}^\alpha \tag{B.11}$$

Then the mass-flux vector can be rewritten as: ⟨AFMS⟩

$$\mathbf{j}_\alpha = \frac{\zeta}{m}[\mathbf{G}^\alpha - \nabla \cdot \boldsymbol{\pi}^{(\phi)\alpha} - \nabla \cdot \boldsymbol{\pi}^{(e)\alpha} - \mathbf{S}^{(b)\alpha}] \tag{B.12}$$

Next we show that $\mathbf{S}^{(b)\alpha}$ is the same as $[\nabla \cdot \boldsymbol{\pi}^{(k)}]$. This requires the use of Eq. (11.5) for the Brownian forces:

$$\begin{aligned}
\mathbf{S}^{(b)\alpha} &= -\Sigma_\nu \int \mathbf{F}_\nu^{(b)\alpha}\, \bar{\Psi}_\alpha(\mathbf{r}_\nu^\alpha, t)\, \delta(\mathbf{r}_\nu^\alpha - \mathbf{r})\, d\mathbf{r}^\alpha \\
&= \Sigma_{\nu\mu} \int \left(\frac{\partial}{\partial \mathbf{r}_\nu^\alpha} \cdot m_\nu^\alpha [[(\dot{\mathbf{r}}_\mu^\alpha - \mathbf{u}_\mu^\alpha)(\dot{\mathbf{r}}_\nu^\alpha - \mathbf{u}_\nu^\alpha)]]^\alpha \bar{\Psi}_\alpha \right) \delta(\mathbf{r}_\nu^\alpha - \mathbf{r})\, d\mathbf{r}^\alpha \\
&= -\Sigma_{\nu\mu} \int m_\nu^\alpha [[(\dot{\mathbf{r}}_\nu^\alpha - \mathbf{u}_\nu^\alpha)(\dot{\mathbf{r}}_\mu^\alpha - \mathbf{u}_\mu^\alpha)]]^\alpha \bar{\Psi}_\alpha \cdot \frac{\partial}{\partial \mathbf{r}_\mu^\alpha} \delta(\mathbf{r}_\nu^\alpha - \mathbf{r})\, d\mathbf{r}^\alpha \\
&= \nabla \cdot \Sigma_\nu \int m_\nu^\alpha [[(\dot{\mathbf{r}}_\nu^\alpha - \mathbf{u}_\nu^\alpha)(\dot{\mathbf{r}}_\nu^\alpha - \mathbf{u}_\nu^\alpha)]]^\alpha \bar{\Psi}_\alpha \delta(\mathbf{r}_\nu^\alpha - \mathbf{r})\, d\mathbf{r}^\alpha
\end{aligned} \tag{B.13}$$

Next we examine one term in the first ⟨···⟩ expression in Eq. (7.3) by using Eq. (5.7) this gives:

$$\begin{aligned}
&\frac{1}{m} \Sigma_\nu \int [[\mathbf{p}_\nu^\alpha \mathbf{p}_\nu^\alpha]]^\alpha \bar{\Psi}_\alpha(\mathbf{r}_\nu^\alpha, t)\, \delta(\mathbf{r}_\nu^\alpha - \mathbf{r})\, d\mathbf{r}^\alpha \\
&\quad = m\Sigma_\nu \int [[(\dot{\mathbf{r}}_\nu^\alpha - \mathbf{v})(\dot{\mathbf{r}}_\nu^\alpha - \mathbf{v})]]^\alpha \bar{\Psi}_\alpha(\mathbf{r}_\nu^\alpha, t)\, \delta(\mathbf{r}_\nu^\alpha - \mathbf{r})\, d\mathbf{r}^\alpha
\end{aligned}$$

$$+ m\Sigma_\nu \int [[\dot{\mathbf{r}}_\nu^\alpha \mathbf{v} + \mathbf{v}\dot{\mathbf{r}}_\nu^\alpha - \mathbf{v}\mathbf{v})]]^\alpha \bar{\Psi}_\alpha(\mathbf{r}_\nu^\alpha, t)\,\delta(\mathbf{r}_\nu^\alpha - \mathbf{r})\,d\mathbf{r}^\alpha$$

$$= \boldsymbol{\pi}^{(k)\alpha} + \rho_\alpha \mathbf{v}\mathbf{v} \tag{B.14}$$

Then we rewrite $\boldsymbol{\pi}^{(k)\alpha}$ following the procedure at the beginning of Sect. 14(a):

$$\boldsymbol{\pi}^{(k)\alpha} \doteq m\Sigma_\nu \int [[(\dot{\mathbf{r}}_\nu^\alpha - \mathbf{u}_\nu^\alpha)(\dot{\mathbf{r}}_\nu^\alpha - \mathbf{u}_\nu^\alpha)]]^\alpha \bar{\Psi}_\alpha(\mathbf{r}_\nu^\alpha, t)\,\delta(\mathbf{r}_\nu^\alpha - \mathbf{r})\,d\mathbf{r}^\alpha \tag{B.15}$$

A second term is omitted in as much as it is presumably small (see Sect. 14.1.1). We can conclude, therefore, from Eqs. (B.12), (B.13), and (B.15) that: ⟨AFMS⟩

$$\mathbf{j}_\alpha = \frac{m}{\zeta}[\mathbf{G}^\alpha - \nabla\cdot\boldsymbol{\pi}^{(\phi)\alpha} - \nabla\cdot\boldsymbol{\pi}^{(e)\alpha} - \nabla\cdot\boldsymbol{\pi}^{(k)\alpha}] \tag{B.16}$$

or ⟨AFMS⟩

$$\mathbf{j}_\alpha = \frac{m}{\zeta}[\mathbf{G}^\alpha - \nabla\cdot\boldsymbol{\pi}_\alpha] \tag{B.17}$$

This result is in exact agreement with the continuum mechanics results of Öttinger [29], provided that his n is interpreted as our $\rho_\alpha/N_\alpha m$. Furthermore it reduces to the result of Beris and Mavrantzas [30] for Rouse chains. If Eq. (B.17) is summed on α, the left side gives zero, but the right side does not. It can be shown that this is a result of assumptions A and F. If assumption A is avoided and assumption F is replaced by an empiricism for $\mathbf{F}_{\nu\mu}^{(d)\alpha\beta}$ in Eq. (11.6) that accounts for the different interactions between pairs of chemical species, then Eq. (B.17) is replaced by a set of equations of the Maxwell–Stefan form, and the summation on α leads to an equality [20d].

If the external force is independent of position, and if the fluid is regarded as incompressible, then substitution of the mass flux in Eq. (B.17) into the equation of continuity of Eq. (6.1) gives the *diffusion equation*: ⟨FMS⟩

$$\frac{D\rho_\alpha}{Dt} = \frac{m}{\zeta}\nabla\nabla : \boldsymbol{\pi}_\alpha \tag{B.18}$$

This equation has been derived with very few assumptions.

We now want to consider this equation further when the external force contribution and the intermolecular contributions to the stress tensor may be neglected, that is when $\boldsymbol{\pi}_\alpha = \boldsymbol{\pi}^{(k)\alpha} + \boldsymbol{\pi}^{(\phi)\alpha}$.

We begin by considering the kinetic contribution given in the first line of Eq. (7.8); we use Eq. (12.20) to evaluate the momentum-space average, and drop the second term, which is believed to be small:

$$\boldsymbol{\pi}^{(k)\alpha} = \Sigma_\nu kT\boldsymbol{\delta}\int \Psi_\alpha(\mathbf{r} - \mathbf{R}_\nu^\alpha, \mathbf{Q}^\alpha, t)\,d\mathbf{Q}^\alpha = kT\boldsymbol{\delta}(\rho_\alpha/m) \tag{B.19}$$

Here we have also used the second line of Eq. (6.7) and it has been assumed that all the beads of the molecules have the same mass.

Next we examine the intramolecular contribution, given by Eq. (7.15), and specialize this for the *Rouse chain model*:

$$\boldsymbol{\pi}^{(\phi)\alpha} = \Sigma_\nu \iint_0^1 \mathbf{R}_\nu^\alpha \mathbf{F}_\nu^{(\phi)\alpha} \Psi_\alpha(\mathbf{r} - \xi \mathbf{R}_\nu^\alpha, \mathbf{Q}^\alpha, t)\, d\xi d\mathbf{Q}^\alpha$$

$$= -H\Sigma_k \iint_0^1 \mathbf{Q}_k^\alpha \mathbf{Q}_k^\alpha \Psi_\alpha(\mathbf{r} - \xi \mathbf{R}_\nu^\alpha, \mathbf{Q}^\alpha, t)\, d\xi d\mathbf{Q}^\alpha \tag{B.20}$$

This quantity, evaluated at equilibrium, is

$$\boldsymbol{\pi}_{eq}^{(\phi)} = -H\Sigma_k \int \mathbf{Q}_k^\alpha \mathbf{Q}_k^\alpha \Psi_{\alpha, eq}\, d\mathbf{Q}^\alpha = -(N_\alpha - 1) n_\alpha kT \boldsymbol{\delta} \tag{B.21}$$

as may be seen from Eq. (14.8) with all the $\boldsymbol{\alpha}_k$ set equal to the unit tensor $\boldsymbol{\delta}$.

We now combine the last four equations to obtain ⟨FMRSX⟩

$$\frac{D\rho_\alpha}{Dt} = \frac{m}{\zeta} \nabla\nabla : \left(\frac{kT\rho_\alpha}{m} \boldsymbol{\delta} + (\boldsymbol{\pi}^{(\phi)\alpha} - \boldsymbol{\pi}_{eq}^{(\phi)\alpha}) - (N_\alpha - 1) n_\alpha kT \boldsymbol{\delta} \right) \tag{B.22}$$

or, for systems with constant temperature: ⟨FMRSXT$_0$⟩

$$\frac{D\rho_\alpha}{Dt} = \frac{kT}{N_\alpha \zeta} \nabla^2 \rho_\alpha + \frac{kT}{N_\alpha \zeta} \nabla\nabla : ((N_\alpha - 1)(\rho_\alpha - m_m^\alpha n_\alpha)\boldsymbol{\delta} + (\boldsymbol{\pi}^{(\phi)\alpha} - \boldsymbol{\pi}_{eq}^{(\phi)\alpha})) \tag{B.23}$$

which is a generalization of *Fick's second law of diffusion*. This equation shows that there is an extra term in the diffusion equation describing the effects of the extension in space of the polymer molecules (the term containing $(\rho_\alpha - m_m^\alpha n_\alpha)$) and the effects of the intramolecular forces within the polymer molecules (the term containing $\boldsymbol{\pi}^{(\phi)\alpha}$). To examine these effects further, we note that:

$$\rho_\alpha - m_m^\alpha n_\alpha = -\nabla \cdot \Sigma_\nu m_\nu^\alpha \iint_0^1 \mathbf{R}_\nu^\alpha \Psi_\alpha(\mathbf{r} - \xi \mathbf{R}_\nu^\alpha, \mathbf{Q}^\alpha, t)\, d\xi d\mathbf{Q}^\alpha$$

$$= \nabla\nabla : \Sigma_\nu m_\nu^\alpha \iint_0^1 \mathbf{R}_\nu^\alpha \mathbf{R}_\nu^\alpha \Psi_\alpha(\mathbf{r} - \xi \mathbf{R}_\nu^\alpha, \mathbf{Q}^\alpha, t)(1 - \xi)\, d\xi d\mathbf{Q}^\alpha \tag{B.24}$$

the second form being obtained by doing a Taylor-series expansion of the distribution function in the first line. Comparison of this expression with that in Eq. (B.20) suggests that the term containing $(\rho_\alpha - m_m^\alpha n_\alpha)$ may be of the same order as that containing $\boldsymbol{\pi}^{(\phi)\alpha}$.

The extra terms in Eq. (B.24) are needed in order to describe the *Uhlenhopp effect* – the migration of polymer molecules in inhomogeneous flow fields (that is, flow fields in which the velocity gradients are functions of position) [34a]. It is evident that if one neglects higher-order terms in Eq. (6.7) and if one uses the expression for $\boldsymbol{\pi}^{(\phi)\alpha}$ in Eq. (14.11) (for uniform velocity gradients), it is not possible to describe the polymer migration using Eq. (B.23). It is probable that, in order to progress further in describing polymer migration in inhomogeneous flows, it will be necessary, in developing Eq. (13.1), to replace $[\boldsymbol{\kappa}\cdot\mathbf{r}]$ by the entire Taylor series $[\boldsymbol{\kappa}\cdot\mathbf{r}] + [\boldsymbol{\Lambda}:\mathbf{rr}] + \cdots$ (where $\boldsymbol{\Lambda}$ is a third-order tensor) and to solve the resulting equation for Ψ_α in Eq. (13.4).

Appendix C
An Equation of Change for the Temperature in Flowing Polymeric Liquids

Thus far, we have given two forms of the energy equation: Eq. (8.1), for the sum of the internal and kinetic energies, and Eq. (8.1a), for the internal energy alone. In order to solve practical problems in nonisothermal flow, we need to have an equation of change for the temperature, the latter being defined in Eq. (12.8) or Eq. (12.9). In this Appendix we show how to transform Eq. (8.1a) into an equation of change for the temperature. It is further demonstrated that, in the limit of no velocity gradients, the Onsager reciprocal relations are satisfied.

C.1 The Internal Energy per Unit Volume for a Solution

The internal energy per unit volume of the liquid, $\rho\hat{U}$, appearing in Eq. (8.1a), can be obtained from Eqs. (8.3) and (8.4):

$$\rho\hat{U} = \left\langle \sum_{\alpha i \nu} \left(\tfrac{1}{2} m_\nu^\alpha \left(\frac{\mathbf{p}_\nu^\alpha}{m_\nu^\alpha} - \mathbf{v} \right)^2 + \tfrac{1}{2} \sum_\mu \phi_{\nu\mu}^{\alpha i} + \tfrac{1}{2} \sum_{\beta j \mu} \phi_{\nu\mu}^{(d)\alpha i, \beta j} \right) \delta(\mathbf{r}_\nu^{\alpha i} - \mathbf{r}) \right\rangle \quad \text{(C.1)}$$

It is now convenient to write this quantity as a sum of contributions involving single molecules (s) plus a sum of contributions involving doublets (d); then the singlet contributions are split further into kinetic (k) and intramolecular (ϕ) parts:

$$\rho\hat{U} = \Sigma_\alpha U_\alpha^{(s)} + \Sigma_\alpha \Sigma_\beta U_{\alpha\beta}^{(d)} = \Sigma_\alpha U_\alpha^{(k)} + \Sigma_\alpha U_\alpha^{(\phi)} + \Sigma_\alpha \Sigma_\beta U_{\alpha\beta}^{(d)} \quad \text{(C.2)}$$

All the U's on the right side have dimensions of energy per unit volume. The individual contributions can then be written in terms of lower-order distribution functions by using Eqs. (5.9) and (5.19):

$$\begin{aligned} U_\alpha^{(k)} &= \sum_\nu \tfrac{1}{2} m_\nu^\alpha \int \left[\!\left[\left(\frac{\mathbf{p}_\nu^\alpha}{m_\nu^\alpha} - \mathbf{v} \right)^2 \right]\!\right]^\alpha \Psi_\alpha(\mathbf{r} - \mathbf{R}_\nu^\alpha, \mathbf{Q}^\alpha, t)\, d\mathbf{Q}^\alpha \\ &= \tfrac{3}{2} n_\alpha kT \sum_\nu \int \psi_\alpha(\mathbf{r} - \mathbf{R}_\nu^\alpha, \mathbf{Q}^\alpha, t)\, d\mathbf{Q}^\alpha \\ &\quad + \sum_\nu \tfrac{1}{2} n_\alpha m_\nu^\alpha \int (\mathbf{u}_\nu^\alpha - \mathbf{v})^2 \psi_\alpha(\mathbf{r} - \mathbf{R}_\nu^\alpha, \mathbf{Q}^\alpha, t)\, d\mathbf{Q}^\alpha \end{aligned} \quad \text{(C.3)}$$

$$U_{\alpha}^{(\phi)} = \tfrac{1}{2} \sum_{\nu\mu} \int \phi_{\nu\mu}^{\alpha} \Psi_{\alpha}(\mathbf{r} - \mathbf{R}_{\nu}^{\alpha}, \mathbf{Q}^{\alpha}, t)\, d\mathbf{Q}^{\alpha} \tag{C.4}$$

$$U_{\alpha\beta}^{(d)} = \tfrac{1}{2} \sum_{\nu\mu} \int \phi_{\nu\mu}^{(d)\alpha\beta}\, \tilde{\Psi}_{\alpha\beta}\left(\mathbf{r} - \mathbf{R}_{\nu}^{\alpha} + \frac{m_m^{\beta}}{m_m^{\alpha} + m_m^{\beta}} \mathbf{R}_{\alpha\beta}, \mathbf{R}_{\alpha\beta}, \mathbf{Q}^{\alpha}, \mathbf{Q}^{\beta}, t\right) \cdot d\mathbf{R}_{\alpha\beta}\, d\mathbf{Q}^{\alpha} d\mathbf{Q}^{\beta} \tag{C.5}$$

The second form in Eq. (C.3) was obtained by using the trace of Eq. (12.20), and the fact that the double-bracket has the same arguments as the distribution function.

We now restrict ourselves to dilute solutions, so that we can neglect all intermolecular interactions except those between pairs of solvent molecules (designated by subscript s). Then we have:

$$\rho \hat{U} = U_s^{(s)} + \Sigma_{\alpha}' U_{\alpha}^{(s)} + U_{ss}^{(d)} = U_s + \Sigma_{\alpha}' U_{\alpha}^{(s)} \tag{C.6}$$

where the primes indicate that the sum is over the solute molecules only, and U_s is the internal energy per unit volume of the pure solute.

Let us now define a "local equilibrium distribution function" thus:

$$\psi_{\alpha, eq}(T) = \frac{\exp(-\Sigma_{\nu\mu} \phi_{\nu\mu}^{\alpha}/2kT)}{\int \exp(-\Sigma_{\nu\mu} \phi_{\nu\mu}^{\alpha}/2kT)\, d\mathbf{Q}^{\alpha}} \tag{C.7}$$

This is a function of the local temperature $T(\mathbf{r}, t)$. With this function we can then define the "local equilibrium internal energy" per unit volume for species α as:

$$U_{\alpha, eq} = U_{\alpha, eq}^{(k)} + U_{\alpha, eq}^{(\phi)} \tag{C.8}$$

where

$$U_{\alpha, eq}^{(k)} = \tfrac{3}{2} N_{\alpha} n_{\alpha} kT \tag{C.9}$$

$$U_{\alpha, eq}^{(\phi)} = \tfrac{1}{2} n_{\alpha} \Sigma_{\nu\mu} \int \phi_{\nu\mu}^{\alpha}\, \psi_{\alpha, eq}(T)\, d\mathbf{Q}^{\alpha} \tag{C.10}$$

We may also define the "local gradient internal energy", which is the part of the internal energy that results from gradients of temperature, concentration, and velocity, as:

$$U_{\alpha, g} = U_{\alpha, g}^{(k)} + U_{\alpha, g}^{(\phi)} \tag{C.11}$$

where

$$U_{\alpha, g}^{(k)} = \tfrac{3}{2} n_{\alpha} kT \Sigma_{\nu} \int [\psi_{\alpha}(\mathbf{r} - \mathbf{R}_{\nu}^{\alpha}, \mathbf{Q}^{\alpha}, t) - \psi_{\alpha, eq}(T)]\, d\mathbf{Q}^{\alpha} + \sum_{\nu} \tfrac{1}{2} n_{\alpha} m_{\nu}^{\alpha} \int (\mathbf{u}_{\nu}^{\alpha} - \mathbf{v})^2 \psi_{\alpha}(\mathbf{r} - \mathbf{R}_{\nu}^{\alpha}, \mathbf{Q}^{\alpha}, t)\, d\mathbf{Q}^{\alpha} \tag{C.12}$$

$$U_{\alpha, g}^{(\phi)} = \tfrac{1}{2} n_{\alpha} \Sigma_{\nu\mu} \int \phi_{\nu\mu}^{\alpha} [\psi_{\alpha}(\mathbf{r} - \mathbf{R}_{\nu}^{\alpha}, \mathbf{Q}^{\alpha}, t) - \psi_{\alpha, eq}(T)]\, d\mathbf{Q}^{\alpha} \tag{C.13}$$

The internal energy per unit volume is then given by:

$$\rho\hat{U} = U_s + \Sigma'_\alpha U_{\alpha,eq} + \Sigma'_\alpha U_{\alpha,g} \tag{C.14}$$

for dilute solutions.

C.2 The Internal Energy Equation in Terms of the Temperature

When the internal energy expression in Eq. (C. 14) is substituted into Eq. (8.1a), we get:

$$\frac{\partial}{\partial t}(U_s + \Sigma'_\alpha U_{\alpha,eq}) + (\nabla\cdot(U_s + \Sigma'_\alpha U_{\alpha,eq})\mathbf{v}) = -(\nabla\cdot\mathbf{q}) - (\boldsymbol{\pi}^\dagger:\nabla\mathbf{v}) + J$$
$$-\frac{\partial}{\partial t}(\Sigma'_\alpha U_{\alpha,g}) - (\nabla\cdot(\Sigma'_\alpha U_{\alpha,g})\mathbf{v}) \tag{C.15}$$

Next we write $U_{\alpha,eq} = n_\alpha u_{\alpha,eq}$, in which n_α is a function of $\mathbf{r}$ and t, and the internal energy per molecule, $u_{\alpha,eq}$, depends on $\mathbf{r}$ and t only through the temperature T. We also approximate U_s as $U_s = n_s u_{s,eq}$, where $u_{s,eq}$ is a function of T alone; this is only approximate, because $U_{ss}^{(d)}$ depends on the density as well as on the gradients of the temperature, concentration, and velocity.

We also define contributions to the heat capacities per molecule:

$$c_{v,s} = du_{s,eq}/dT \quad \text{and} \quad c_{v,\alpha} = du_{\alpha,eq}/dT \tag{C.16}$$

so that the heat capacity of the solution is:

$$C_v = n_s c_{v,s} + \Sigma'_\alpha n_\alpha c_{v,\alpha} = \Sigma_\alpha n_\alpha c_{v,\alpha} \tag{C.17}$$

where the unprimed sum implies a summation over all species including the solvent.

With thcse dcfnitions Eq. (C.15) then becomes for an incompressible fluid: ⟨EIN⟩

$$C_v\frac{DT}{Dt} = -(\nabla\cdot\mathbf{q}) - (\boldsymbol{\pi}^\dagger:\nabla\mathbf{v}) + J - \left(\Sigma_\alpha u_{\alpha,eq}\frac{Dn_\alpha}{Dt}\right) - \frac{D}{Dt}(\Sigma'_\alpha U_{\alpha,g}) \tag{C.18}$$

If we neglect higher terms in the Taylor series expansion in Eq. (6.7) so that $\rho_\alpha = n_\alpha m_m^\alpha$, then the next-to-last term in Eq. (C.18) can be rewritten with the help of the equation of continuity, Eq. (6.1):

$$-\left(\Sigma'_\alpha u_{\alpha,eq}\frac{Dn_\alpha}{Dt}\right) = -\left(\Sigma'_\alpha\frac{u_{\alpha,eq}}{m_m^\alpha}\frac{D\rho_\alpha}{Dt}\right) = +\left(\Sigma'_\alpha\frac{u_{\alpha,eq}}{m_m^\alpha}\nabla\cdot\mathbf{j}_\alpha\right) \tag{C.19}$$

so that the energy equation becomes finally:

$$C_v\frac{DT}{Dt} = -(\nabla\cdot\mathbf{q}) - (\boldsymbol{\pi}^\dagger:\nabla\mathbf{v}) + J + \left(\Sigma_\alpha\frac{u_{\alpha,eq}}{m_m^\alpha}\nabla\cdot\mathbf{j}_\alpha\right) - \frac{D}{Dt}(\Sigma'_\alpha U_{\alpha,g}) \tag{C.20}$$

This equation differs from the usual equation of change for the temperature only in its last term (see Eq. (G) on p. 562 of [11]), appropriately simplified for an incompressible fluid with no chemical reactions. In developing this equation we have used only the local values of temperature and concentration to define the heat capacity and the $u_{\alpha,eq}$ quantities. This is consistent with the normal practice in fluid dynamics in transforming the equation of change for internal energy into an equation of change for temperature. The effect of the gradients on the internal energy of the solution leads to the last term in Eq. (C.20), which contains $U_{\alpha,g}$ and reflects the viscoelastic nature of the fluid.

Therefore we next turn our attention to the quantity $U_{\alpha,g}$ of Eq. (C.11), using only the lowest terms in the Taylor series expansion of the distribution function. The first term of the contribution in Eq. (C.12) vanishes; the second term in that equation is closely related to the trace of $\pi_2^{(k)}(1)$ in Eq. (14.9) and can be considered to be negligible. The contribution in Eq. (C.13) does not vanish, and therefore the lowest-order expression for the gradient internal energy $U_{\alpha,g}$ is:

$$U_{\alpha,g} = \tfrac{1}{2} n_\alpha \Sigma_\nu \Sigma_\mu \int \phi^\alpha_{\nu\mu} [\psi_\alpha(\mathbf{r}, \mathbf{Q}^\alpha, t) - \psi_{\alpha,eq}(T)] \, d\mathbf{Q}^\alpha \tag{C.21}$$

Thus, only the potential energy (from the elastic springs) contributes to $U_{\alpha,g}$, the kinetic energy making no contribution here because of our definition of T in Eq. (12.9) and the use of the Maxwellian velocity distribution in Eq. (12.11). We now develop the integral in Eq. (C.21) for the Rouse chain model.

We note that the integral in Eq. (C.21) is very closely related to the expression for the intramolecular part of the stress tensor given in Eq. (14.10) and (14.11) for the *Rouse chain model*. Specifically we note that:

$$\begin{aligned} tr\boldsymbol{\pi}^{(\phi)\alpha}(1) &= -2 \int \phi^\alpha \Psi_\alpha(\mathbf{r}, \mathbf{Q}^\alpha, t)\, d\mathbf{Q}^\alpha \\ &= -\Sigma_\nu \Sigma_\mu \int \phi^\alpha_{\nu\mu} \Psi_\alpha(\mathbf{r}, \mathbf{Q}^\alpha, t)\, d\mathbf{Q}^\alpha \end{aligned} \tag{C.22}$$

$$tr\boldsymbol{\pi}^{(\phi)\alpha}_{eq}(1) = -3(N_\alpha - 1) n_\alpha kT \tag{C.23}$$

Hence from the last three equations we get:

$$\begin{aligned} U_{\alpha,g} &= -\tfrac{1}{2} tr\boldsymbol{\pi}^{(\phi)\alpha} - \tfrac{3}{2}(N_\alpha - 1) n_\alpha kT \\ &= -\tfrac{1}{2} tr\boldsymbol{\pi}_\alpha + \tfrac{3}{2} N_\alpha n_\alpha kT - \tfrac{3}{2}(N_\alpha - 1) n_\alpha kT \\ &= -\tfrac{1}{2} tr\boldsymbol{\tau}_\alpha \end{aligned} \tag{C.24}$$

Equations (14.7), (14.12), and (14.13) have been used to get the last two forms of the equation. With this expression for $U_{\alpha,g}$ the equation of change for the temperature becomes: ⟨EINR⟩

$$C_v \frac{DT}{Dt} = -(\nabla \cdot \mathbf{q}) - (\boldsymbol{\pi}^\dagger : \nabla \mathbf{v}) + J + \left(\Sigma'_\alpha \frac{u_{\alpha,eq}}{m^\alpha_m} \nabla \cdot \mathbf{j}_\alpha \right) + \frac{D}{Dt} (\Sigma'_\alpha \tfrac{1}{2} tr\boldsymbol{\tau}_\alpha) \tag{C.25}$$

The last term is closely related to the rate of stretching of the springs in the molecular model beyond their thermal equilibrium lengths (DPL, Eq. (13.4–11)). It is thus an "elastic contribution", in addition to that implicit in $\boldsymbol{\pi}^\dagger$, that does not appear in the equation for Newtonian fluids. Such a term has been included in the equation of change for temperature by Wiest [35]. If this last term is omitted, then the usual equation of change for temperature for Newtonian fluid mixtures is obtained ([11], Table 18.3-1).

It should be emphasized that the above development is a lowest-order development in that only the first corrections to the local equilibrium values of the internal energy and heat capacity have been used. This is consistent with current practice in fluid dynamics and nonequilibrium thermodynamics. Further refinements can be developed if it is desired to make contact with some of the recent work in "extended irreversible thermodynamics" [36, 37].

C.3 The Onsager Reciprocal Relations

According to Eqs. (15.6) and (16.31), for a solution with one polymer solute (modeled as *Hookean dumbbells*) which is devoid of velocity gradients:

$$\mathbf{j}_\alpha = -\frac{m_m^\alpha n_\alpha kT}{\zeta}\left(\tfrac{1}{2}\mathbf{b}_\alpha + \tfrac{1}{2}\mathbf{a}\right) \tag{C.26}$$

$$\mathbf{q}_\alpha = -\frac{n_\alpha k^2 T^2}{\zeta}\left(\tfrac{9}{2}\mathbf{b}_\alpha + \tfrac{11}{4}\mathbf{a}\right) \tag{C.27}$$

For Hookean dumbbells, we find from Eqs. (C.9) and (C.10) that:

$$u_{\alpha,eq}^{(k)} = 3kT \tag{C.28}$$

$$u_{\alpha,eq}^{(\phi)} = \tfrac{1}{2} H \int Q^2 \psi_{\alpha,eq} d\mathbf{Q} = \tfrac{3}{2} kT \tag{C.29}$$

The sum of these gives the local equilibrium internal energy for the solute species, which is $u_{\alpha,eq} = \frac{9}{2}kT$.

We now rewrite Eqs. (C.26) and (C.27) in the form required by the thermodynamics of irreversible processes (see Chapter 6 of [30a] and Chapter 11 of [12]):

$$\mathbf{j}_\alpha = -\frac{1}{2}\frac{m_m^\alpha kT}{\zeta}\nabla n_\alpha - \frac{1}{2}\frac{m_m^\alpha n_\alpha k}{\zeta}\nabla T \tag{C.30}$$

$$\mathbf{q}_\alpha - \frac{9kT}{2m_m^\alpha}\mathbf{j}_\alpha = -\frac{1}{2}\frac{k^2T^2}{\zeta}\nabla n_\alpha - \frac{9}{4}\frac{n_\alpha k^2 T}{\zeta}\nabla T \tag{C.31}$$

The term in the mass flux involving the temperature gradient describes the *Soret* (or *thermal-diffusion*) *effect*; the term on the right side of Eq. (31) involving the concentration gradient describes the *Dufour* (or *diffusion-thermo*) *effect*.

When the fluxes are taken to be $\mathbf{j}_\alpha$ and $\mathbf{q}_\alpha - (9kT/2m_m^\alpha)\mathbf{j}_\alpha$, as suggested by the equation of change for entropy, then the numerical coefficients in the cross terms are the same, in agreement with the Onsager reciprocal relations. This provides one check on the evaluation of the flux expressions for the Hookean dumbbell model.

References

1. Kramers HA (1944) Physica 11: 1–19
2. Kirkwood JG (1967) Macromolecules. Gordon and Breach, New York
3. Rouse PE Jr (1953) J Chem Phys 21: 1272–1280
4. Zimm BH (1956) J Chem Phys 24: 269–278
5. Curtiss CF, Bird RB, Hassager O (1976) Adv Chem Phys 35: 31–117
6. Irving JH, Kirkwood JG (1950) J Chem Phys 18: 817–829
7. Bird RB, Hassager O, Armstrong RC, Curtiss CF (1977) Dynamics of polymeric liquids, Vol. 2. Wiley, New York
8. Bird RB, Armstrong RC, Hassager O (1987) Dynamics of polymeric liquids, vol 1, 2nd edn, Wiley-Interscience, New York; Bird RB, Curtiss CF, Armstrong RC, Hassager O (1987) Dynamics of polymeric liquids, vol 2, 2nd edn. Wiley-Interscience, New York
9. Curtiss CF, Bird RB (1982) J Chem Phys 74: 2016–2033; Bird RB, Saab HH, Curtiss CF (1982) J Phys Chem 86: 1102–1106 (1982) and J Chem Phys 4747–4757; Saab HH, Bird RB, Curtiss CF (1982) J Chem Phys 77: 4758–4766; Fan XJ, Bird RB (1984) J Non-Newtonian Fluid Mech. 15: 341–373; Curtiss CF, Bird RB (1983) Physica 118A: 191–204
10. Schieber JD, Curtiss CF, Bird RB (1986) Ind Eng Chem Fundamentals 25: 471–475; Schieber JD (1987) J Chem Phys 87: 4917–4936; Lodge AS, Schieber JD, Bird RB (1988) 88 4001–4007

10a. Hassager O (1974) J Chem Phys 60: 2111–2124

10b. Gottlieb M, Bird RB (1976) J Chem Phys 65: 2467–2468; Gottlieb M (1977) Computers in Chem 1: 155–160

10c. Van Kampen N (1981) Appl Sci Res 37: 67–75; for a quantum mechanical treatment see Bruch LW, Goebel CJ (1981) J Chem Phys 74: 4040–4047

10d. Bird RB, Öttinger HC (1992) Ann Rev Phys Chem 43: 371–406

11. Bird RB, Stewart WE, Lightfoot EN (1960) Transport phenomena Wiley, New York

11a. Dahler JS, Scriven LE (1961) Nature 192: 36–37 (No. 4797); Dahler JS (1965) chap 15 In: Seeger RJ, Temple G (eds) Research frontiers in fluid dynamics. Wiley-Interscience, New York

12. Hirschfelder JO, Curtiss CF, Bird RB (1964) Molecular theory of gases and liquids. Wiley, New York [Second printing with corrections and added notes]
13. Curtiss CF (1974) Physical chemistry. Vol VIA, Eyring H, Henderson D, Jost W (eds) Academic Press, New York
14. Curtiss CF (1992) Theor Chim Acta 82: 75–91

14a. Doi M, Edwards SF (1986) Theory of Polymer Dynamics, Oxford Univ Press

15. Fixman M (1965) J Chem Phys 42: 3831–3837
16. Schieber JD, Öttinger HC (1988) J Chem Phys 89: 6972–6981
17. Murphy TJ, Aguirre JL (1972) J Chem Phys 57: 2098–2104
18. Szu SC, Hermans JJ (1974) J Polymer Sci 12: 1743–1751
19. Curtiss CF (1989) Momentum equilibration in polymeric systems. WIS-TCI-746
20. Kuhn W (1933) Koll Zeits 62: 269–285

20a. Giesekus H (1966) Rheol Acta, 5: 26–36; (1982) J Non-Newtonian Fluid Mech. 11: 69–109

20b. Bird RB, DeAguiar JR (1983) J Non-Newtonian Fluid Mech 13: 149–160; DeAguiar JR (1983), J Non-Newtonian Fluid Mech 13: 161–169

20c. Bird RB, Wiest JM (1985) 29: 519–532

20d. Curtiss CF, Bird RB (1996) Proc Nat Acad Sci 93

21. Lodge AS, Wu Y (1971) Rheol Acta 10: 539–553
22. van Wiechen PH, Booij HC (1971) J Eng Math 5: 89–98

23. Öttinger HC (1996) Stochastic processes in polymeric fluids. Springer, Berlin
23a. Osaki K, Schrag JL (1971) Polymer Journal 2: 541–549; Osaki K, Schrag JL, Ferry JD (1972) Macromolecules 5: 144–147
23b. Lodge TP (1993) J Phys Chem 97: 1480–1487; Schrag JL, Stokich TM, Strand DA, Merchak PA, Landry CJT, Radtke DR, Man VF, Lodge TP, Morris RL, Hermann KC, Amelar S, Eastman CE, Smeltzly MA (1991) J Non-Crystalline Solids 131–133: 537–543; Stokich TM, Merchak PA, Radtke DR, White CC, Woltman GR, Schrag JL (1994) Polymer Preprints 35: 138–139
23c. Xu Z, de Pablo JJ, manuscript in preparation
24. Bird RB, Fan, XJ, Curtiss CF (1984) J Non-Newtonian Fluid Mechanics. 15: 85–92
25. Lodge AS (1989) J Rheol 32: 93–95; (1989) Rheol Acta 28: 351–362
26. Bhave AV, Armstrong RC, Brown RA (1991) J Chem Phys 95: 2988–3000
27. El-Kareh AW, Leal LG (1989) J Non-Newtonian Fluid Mechanics, 33: 257–287
28. Öttinger HC (1987) J Chem Phys 87: 3156–3165, 6185–6190; (1989) AIChE Journal 35: 279–286
29. Öttinger HC (1992) Rheol Acta 31: 14–21
30. Beris AN, Mavrantzas VG (1994) J Rheol, 38: 1235–1250; Eq. 29
30a. Landau L, Lifshitz EM (1959) Fluid Mechanics, Addison-Wesley, Reading, Mass (Chapter 6)
31. van den Brule BHAA (1989) Rheol Acta 28: 257–266; (1990) 29: 416–422; (1991) A Contribution to the Micro-Rheological Modeling of Transport Properties. Doctoral Dissertation (Twente University)
31a. Bird RB, Curtiss CF (1996) Rheol Acta 35
32. Piraux L, Ducarme E, Issi J-P, Begin D, Billaud D (1954) Synthetic Metals, 41–43: 129–132
33. Poulaert B, Chielens J-C, Vandenhaende C, Issi J-P, Legras R (1990) Polymer Comm 31: 148–151
33a. Sammler RL, Schrag JL (1988) Macromolecules 21: 1132–1140, 3273–3285; (1989) Macromolecules 22: 3435–3442
33b. Soli AL, Schrag JL (1979) Macromolecules 12: 1159–1163
33c. Martel CJT, Lodge TP, Dibbs MG, Stokich TM, Sammler RL, Carriere CJ, and Schrag JL (1983) Faraday Symp Chem Soc 18: 173–188
33d. Borgbjerg U, de Pablo JJ, Öttinger HC (1994) J Chem Phys 101: 7144–7152 Borgbjerg U, de Pablo JJ (1995) Macromolecules 28: 4540–4547
33e. Carl W, de Pablo JJ (1994) Macromolecular Chem, Theory and Simulations 3: 177–184
33f. Xu Z, de Pablo JJ, Kim S (1994) J Chem Phys 101: 5293–5844; (1995) J Chem Phys 102: 5836–5304
34. Parker EN (1954) Phys Rev. 96: 1686–1689
34a. Shafer RH, Larkin N, Zimm BH (1974) Biophys Chem 2: 180–184; Shafer RH (1974) Biophys. Chem 2: 185–188; Bird RB (1979) J Non-Newtonian Fluid Mech 5: 1–12
35. Wiest JM (1995) Chem Engr Sci (in press)
36. Casas-Vásquez J, Jou D (eds) (1991) Rheological Modelling: Thermodynamical and statistical approaches, Springer, Berlin Heidelberg New York.
37. Beris AN, Edwards BJ (1994) Thermodynamics of Flowing Systems, Oxford University Press

Editor: Prof. J.L. Schrag
Received: 1995

Kinetics of Deformation and Relaxation in Highly Oriented Polymers

S.V. Bronnikov, V.I. Vettegren*, and S.Y. Frenkel
Institute of Macromolecular Compounds, Russian Academy of Sciences, 199004 St. Petersburg/Russia
*A.F. Ioffe Physical Technical Institute, Russian Academy of Sciences, 194021 St. Petersburg/Russia

In this paper, general principles of physical kinetics are used for the description of creep, relaxation of stress and Young's modulus, and fracture of a special group of polymers. The rates of change of the mechanical properties as a function of temperature and time, for stressed or strained highly oriented polymers, is described by Arrhenius type equations. The kinetics of the above-mentioned processes is found to be determined by the probability of formation of excited chemical bonds in macromolecules. The statistics of certain modes of the fundamental vibrations of macromolecules influence the kinetics of their formation decisively. If the quantum statistics of fundamental vibrations is taken into account, an Arrhenius type equation adequately describes the changes in the kinetics of deformation and fracture over a wide temperature range. Relaxation transitions in the polymers studied are explained by the substitution of classical statistics by quantum statistics of the fundamental vibrations.

Advances in Polymer Science, Vol. 125

0 List of Symbols

All necessary and important symbols used in this paper are listed below. The defining equations are also indicated where appropriate.

c_d	concentration of excited bonds (Eq. (13))
c_0	concentration of helices in the macromolecule (Eq. (17))
$\dot{c}_d$	rate of accumulation of excited bonds (Eq. (17))
C	heat capacity (Eq. (45))
C_m	mode heat capacity (Eq. (44))
e	thermal expansion coefficient (Eqs. (33), (43))
E	Young's modulus
F	free energy (Eq. (39))
$F(\theta/T)$	quantum function (Eqs. (33), (36), (45))
G	thermodynamic Grüneisen parameter (Eq. (45))
G_b, G_m, G_{mn}, G_t	mode Grüneisen parameters (Eqs. (40), (42), (55))
G_i	Grüneisen parameter for a regularity band (Eq. (11))
h	Plank's constant
I_A	band intensity in an absorption spectrum (Eq. (14))
I_{AS}	band intensity in the anti-Stokes region (Eq. (15))
I_d	satellite intensity (Eq. (13))
I_{EM}	band intensity in an emission spectrum (Eq. (14))
I_S	band intensity in the Stokes region (Eq. (15))
I_0	main band intensity (Eq. (13))
k	Boltzmann constant
k	rate constant (Eqs. (1), (32))
k_0	pre-exponential (frequency) factor (Eqs. (1), (32))
N	phonon occupation number (Eqs. (14)–(16))
S	isothermal compliance (Eq. (40))
t	time
T	absolute temperature
T_b, T_t	characteristic temperatures (Eqs. (23), (25))
T_{ph}	local phonon temperature (Eq. (16))
T_1, T_2	initial and final temperatures (Eqs. (27), (28))
T_t^*, T_b^*	extrapolated temperatures (Eqs. (23), (25))
U	activation energy depending on stress (Eqs. (1), (37))
U_{fl}	activation energy of fluctuations (Eq. (10)
U_L	energy of atoms interaction (Eq. (39))
U_0	activation energy
V	cell volume (Eq. (40))
W	energy of atomic vibrations (Eq. (29))
W_m	energy of *m*-th vibrational mode (Eq. (30))

Greek Symbols

γ	activation volume

Δ	difference (increase)
$\Delta\nu$	frequency shift in IR spectrum (Eq. (11))
ε	deformation
ε_d	deformation of excited bonds (Eqs. (12), (19))
ε_T	deformation caused by temperature (Eqs. (35), (40))
ε_0	constant deformation
ε_*	rupture deformation of interatomic bonds (Eq. (19))
$\dot{\varepsilon}$	deformation rate
$\dot{\varepsilon}_0$	pre-exponential (frequency) factor (Eqs. (2), (5), (7))
$\dot{\varepsilon}_1, \dot{\varepsilon}_2$	initial and final creep rate (Eqs. (27), (28))
θ	thermodynamic Debye temperature
θ_m	mode Debye temperature (Eq. (31))
ν	frequency of atomic vibrations
$\nu(\varepsilon)$	frequency of atomic vibrations for deformed helices in a macromolecule (Eq. (11))
$\nu(0)$	frequency of atomic vibrations for non-deformed helices in a macromolecule (Eq. (11))
ν_{mn}	frequency of mode vibrations (Eqs. (30), (39), (42))
σ	stress
σ_1, σ_2	initial and final stress (Eqs. (27), (28))
τ_d	average time for generation of the excited bonds (Eqs. (17), (18))
τ_f	time to fracture (Eq. (8))
τ_{fl}	average time between subsequent fluctuations (Eq. (10))
τ_0	pre-exponential (frequency) factor (Eqs. (8), (10), (18))

Subscripts

ap	apparent activation parameters
b	mode of the deformation vibration
d	dilated excited interatomic bonds
e	new ends of macromolecules
f	fracture
m	number of modes of fundamental vibrations
max	maximum frequency of the vibrations
n	number of vibrations in *m*-th mode
p	quasi-elastic deformation (decelerating creep or relaxation of Young's modulus)
s	steady state creep
t	mode of the twisting vibration
tr	true activation parameters

1 Introduction

Physical kinetics applies to physical systems whose elements are far from equilibrium. It is the macroscopic description of the processes which occur in non-equilibrium systems. In physical kinetics, the changes in energy, momentum, and mass transfer in the various physical systems are investigated, as well as the influence of external fields on these systems.

The basic expression of physical kinetics is the Arrhenius equation. In 1889 Arrhenius suggested [1] that the rate of chemical reaction is controlled by the rate constant k:

$$k = k_0 \exp\left(-\frac{U}{\mathbf{k}T}\right) \tag{1}$$

where k_0 is a frequency factor, U is activation energy – which is equal to the barrier height to be overcome to activate the process, **k** is the Boltzmann constant, and T is the absolute temperature.

Becker was the first to suggest, in 1925, [2] that the deformation rate could be expressed by an Arrhenius-type equation. His concept was developed from the recognition of an identity between deformation and chemical reactions. The rate theory of thermally activated processes and, in general, the principles of deformation and fracture were established by Eyring [3], Orowan [4], and J. Frenkel [5]. In the latter part of the 20th century, these principles were further developed for polymers by Krausz and Eyring [6], Dorn [7], Zhurkov [8] and his collaborators [9], Kausch [10], and many others [11–14].

Nowadays, the kinetics of polymer deformation is often supposed to be specific for the different stages in the evolution of deformation. For instance, this seems to be the case for three types of the kinetics of creep: decelerating, steady state, and accelerating creep [6].

When the deformation is too small and its rate is high (e.g. by the action of low amplitude vibrations), a polymer deforms elastically and this process can be described by the Hooke's law [6].

An increase in both deformation and loading time evokes another – viscoelastic or relaxational – mechanism. In this case, a macromolecule is treated as a set of independent relaxors (segments) [13, 14]. The latter change their spatial state by overcoming the potential energy barrier under the action of thermal fluctuations. At a certain temperature, only a finite fraction of the relaxors is "unfrozen", whereas the others remain "frozen". In such a model, the activation energy of creep is treated in terms of the height of the potential barriers to be overcome by relaxators to cause a local deformation. Therefore, the activation energy increases stepwise with increasing temperature [13, 14].

At last, at a final deformation stage, the defects caused by local fracture should be taken into account. At this stage, it will be supposed that the deformation rate is controlled by the rate of submicrocrack generation [9, 10, 15].

During the last 10 years, in this respect rather contradictory experimental data are accumulated.

(i) The Young's modulus of polymers was revealed to depend both on time and temperature. Hence, Hooke's law is not suitable even for small deformations of polymers [16–22].
(ii) The activation energy of deformation, particularly for orientation polymers, seems to be constant, i.e. it does not depend on the extent of deformation [19, 20, 23, 24].

Research methods were developed to observe the excited interatomic bonds in macromolecules. Their relaxation is related to the generation of the elementary deformation and/or to fracture nucleation. The rate of deformation was shown to be controlled by the formation of the excited bonds rather than by their relaxation [24–39].

Finally, the change in the statistics of certain modes of atomic vibrations was revealed to be the implicit cause of the relaxation transitions [34, 35, 40].

These "out-of-the-way" data induce us to reconsider the recent investigations in the field of kinetics of polymer deformation. Mainly, this paper reviews previously published research by the authors and, briefly, by other investigators. Therefore, it is partially controversial in character, in that it reflects the authors' opinion.

2 Abbreviations of Polymer Names

The polymers treated are abbreviated in the paper as follows

Flexible-chain polymers

Nylon 6	poly (ω-caproamide)
PAN	polyacrylonitrile
PBP	polybutene-phtalate
PE	polyethylene
PEO	polyethylene oxyde
PETP	polyethylene terephtalate
POM	polyoxymethylene
PP	polypropylene
PTFE	polytetrafluorethylene
PVA	polyvinyl alcohol

Polyimides

BPT B

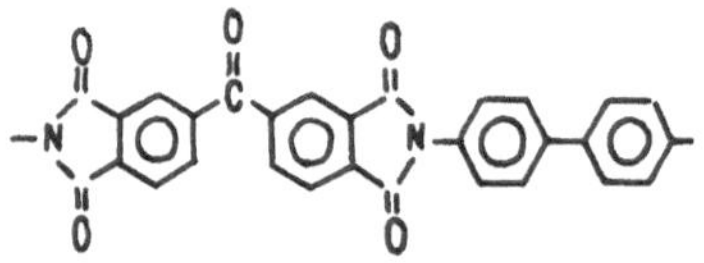

DPO B

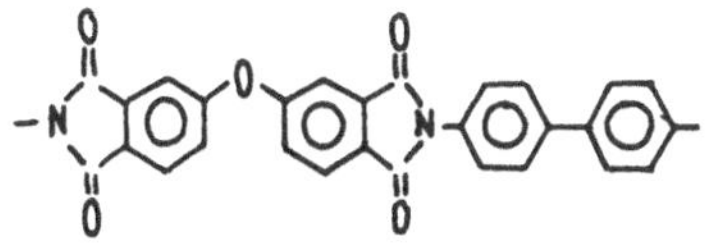

DPO P

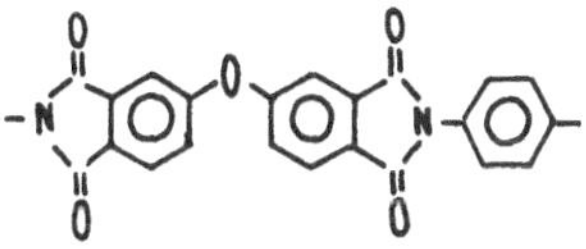

DPT B

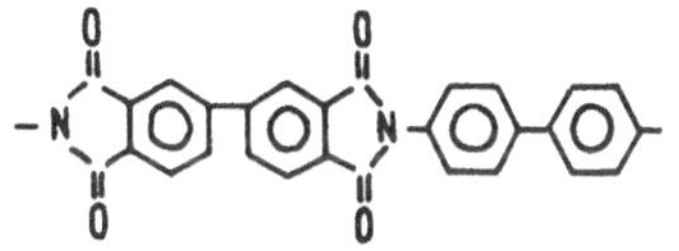

PM

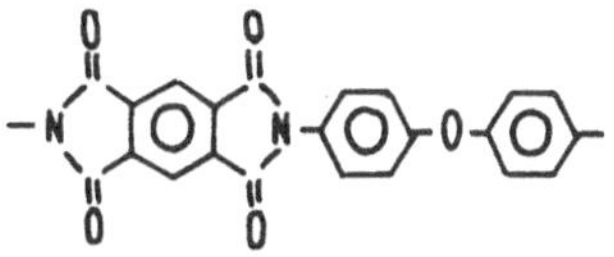

PM S

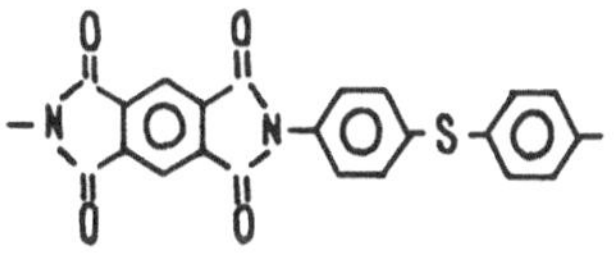

3 Kinetics of Deformation, Relaxation, and Fracture Processes in Polymers in the Temperature Range Between the Main Relaxation Transitions

In this Section, the process of deformation, relaxation, and fracture are examined only within a restricted temperature range between the main (β and α) relaxational transitions, $T_\beta < T < T_\alpha$. The kinetics of creep, relaxation of stress and Young's modulus, and fracture are investigated experimentally as a function of the external stress applied to a sample and/or the increase in temperature. It is shown that the kinetics of the processes considered are described by Arrhenius-type equations. Then, the activation parameters (the energy and the volume) of the kinetic equations are calculated and compared with each other. This procedure demonstrates the identical physical nature of these processes.

Objects for investigation are highly oriented flexible- and rigid-chain polymers in the form of films and fibers. In such samples, macromolecules are packed quite perfectly along the draw axis, and the most extended conformations are realized.

3.1 Kinetics of Decelerating Creep

The first process under investigation is creep, i.e. the evolution of relative deformation with time at a dead load. The general form of a creep curve is shown in Fig. 1. It consists of one, two, or three stages; the existence and length of these stages depend on temperature and/or applied stress [6]. Each stage of creep will be examined separately.

(i) In the first stage, the creep rate decreases gradually with time. This is primary or decelerating creep.
(ii) As a rule, though not always, the decelerating rate tends to achieve a constant value. This is steady-state creep. The creep curve in this stage can be approximated by a straight line, within experimental errors.
(iii) In the tertiary stage, accelerating creep, the creep rate increases as a result of sample weakening (e.g. accumulation of defects such as microcracks) and/or increasing stress. This stage leads to fracture. Accelerating creep covers the shortest time interval; therefore, only its final stage, fracture, is usually investigated.

Let us consider primary creep and try to obtain an empirical equation for the creep rate, $\dot{\varepsilon}_p$, as a function of temperature, T, and applied stress, σ. The results of systematic and careful investigations are presented in Figs. 2 and 3.

Fig. 2 shows that in the plot of $\log \dot{\varepsilon}_p$ versus T^{-1}, the experimental points fall on a "fan" of straight lines at the different σ values. They radiate from an origin on the $\log \dot{\varepsilon}_{0p}$ ordinate.

On the other hand, the experimental points in the plot $\log \dot{\varepsilon}_p$ versus σ (Fig. 3) also fall on a "fan" of straight lines at the different T values. They tend to the same origin, $\log \dot{\varepsilon}_{0p}$.

Combination of these experimental data gives [24, 41, 42]:

$$\ln \dot{\varepsilon}_p = \ln \dot{\varepsilon}_{0p} - \frac{U_{0p} - \gamma_p \sigma}{\mathbf{k}T} . \tag{2}$$

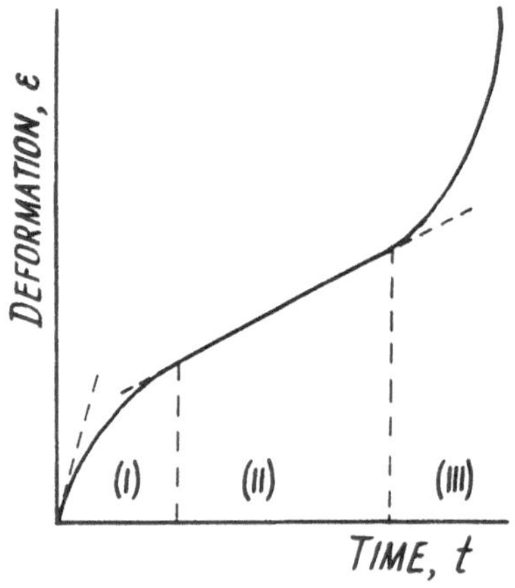

Fig. 1. A creep curve: (i) decelerating (or primary) creep, (ii) steady-state creep, and (iii) accelerating creep

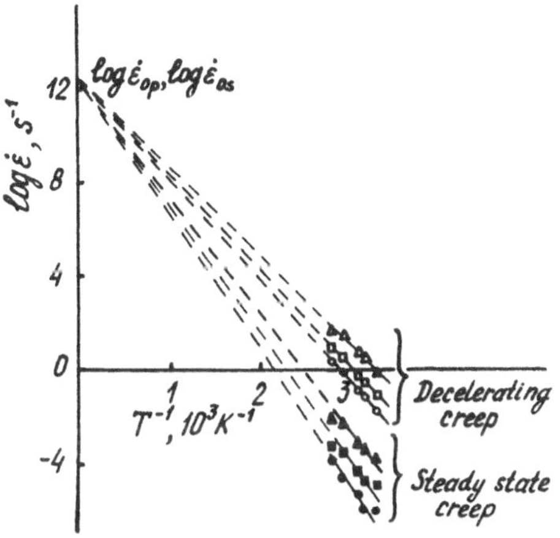

Fig. 2. Log creep rate, $\dot{\varepsilon}$, versus inverse temperature, T^{-1}, for PETP fiber. Applied stress (MPa): (○, ●), 65; (□, ■) 255; and (△, ▲), 550

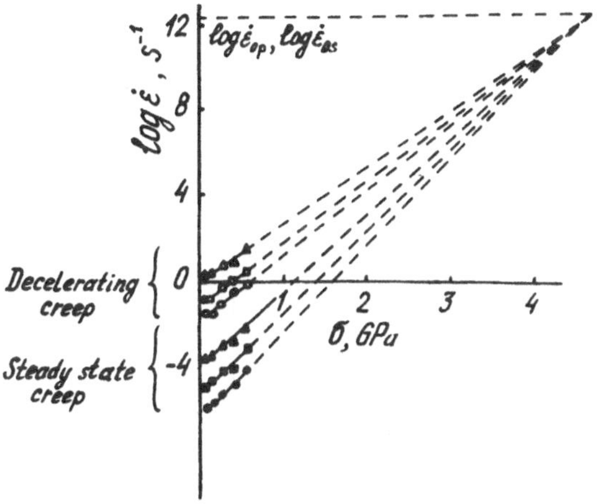

Fig. 3. Log creep rate, $\dot{\varepsilon}$, versus applied stress, σ, for PETP fiber. Temperature (K): (○, ●), 300; (□, ■), 323; and (△, ▲), 353

Equation (2) may by rewritten in the form

$$\dot{\varepsilon}_p = \dot{\varepsilon}_{0p} \exp\left(-\frac{U_{0p} - \gamma_p \sigma}{kT}\right). \tag{3}$$

In Eqs. (2) and (3), $\dot{\varepsilon}_{0p}$ is the ultimate rate at $T \to \infty$, U_{0p} is the activation energy of creep, and γ_p is its activation volume. The dependences presented in Figs. 2 and 3 are typical for all polymers under investigation.

The calculated activation parameters are listed in Table 1. The U_{0p} parameters are equal to some hundreds of kJ mol^{-1} and the γ_p parameters are of the order of 10^{-29} m^3. The evaluated pre-exponential factor, $\dot{\varepsilon}_{0p}$, approximately coincides with the period of atomic vibrations: $\dot{\varepsilon}_{0p} \cong 10^{12}$ Hz.

Table 1. The calculated activation parameters of the kinetic equations

Polymer	Decelerating creep (Eq. (3))		Relaxation of Young's modulus (Eq. (5))		Steady-state creep (Eq. (7))		Fracture ((Eq. 8))		Generation of excited bonds (Eq. (18))		True activation parameters (Eqs. (58), (59))	
	U_{0p} (kJ mol^{-1})	$\gamma_p = \gamma_0\varepsilon_0$ [10^{-29} m^3]	U_{0p} (kJ mol^{-1})	$\gamma_p = \gamma_0\varepsilon_0$ [10^{-29} m^3]	U_{0s} (kJ mol^{-1})	$\gamma_s = \gamma_0$, [10^{-29} m^3]	U_{0f} (kJ mol^{-1})	$\gamma_f = \gamma_0$ [10^{-29} m^3]	U_{0d} (kJ mol^{-1})	$\gamma_d = \gamma_0$ [10^{-29} m^3]	U_{0tr} (kJ mol^{-1})	γ_{tr} [10^{-29} m^3]
Nylon 6	180	50	180	56	160	10	160	18	175	46	130	8.3
PE	220	14	220	14	180	8.1	210	11	–	–	115	4.3
PETP	140	53	140	75	160	12	160	17	160	40	130	1.0
PP	140	9	140	15	140	4.4	140	7	125	10	110	1.0
PVA	130	20	100	17	130	8.6	110	10	–	–	100	1.5
BPT B	230	0.43	210	0.22	220	20	220	17	210	–	170	7.8
DPO B	240	0.28	220	0.21	230	29	240	19	–	–	190	4.8
DPO P	230	0.30	220	0.28	220	11	220	13	–	–	170	4.7
DPT B	220	0.40	210	0.42	220	12	190	14	–	–	170	7.2
PM	200	2.5	220	2.0	220	24	200	32	160	92	170	15.5
PM S	210	9.9	200	8.7	220	62	190	80	–	–	170	20.0

Therefore, the kinetics of decelerating creep for oriented polymers in the range between the main relaxation transitions can be described by an Arrhenius type equation in the form of Eq. (3).

The results of several investigations of primary creep have been published previously. The most careful data in this area have been obtained by Stepanov, Peschanskaya, and Shpeizman [13] and by Bershtein et al. [14]. For this purpose they have used a laser interferometer, which gives high accuracy: the minimum deformation at which the creep rate may be evaluated is 1.5 10^{-4} mm. They have investigated a number of solids, including polymers. The latter were mainly in the isotropic state. Therefore, they have observed and identified a large numbers of relaxation transitions. One should also mention the investigations by Korsukov et al. [43]. Their results also confirm the correctness of Eq. (3).

An other characteristics of the region of small deformations is Young's modulus. The next Section will therefore be devoted to investigations on the relaxation of Young's modulus.

3.2 Kinetics of Relaxation of Young's Modulus and Stress

According to Hooke's law, Young's modulus E, is the coefficient between a small deformation, ε_p, and stress, σ: $\sigma = E\varepsilon_p$. However, the experimental measurements of Young's modulus show that it is not a constant parameter, as assumed in the classic theory of elasticity: it depends on both temperature and time scale. It is known that an increase in the deformation rate (or deformation frequency) causes an increase in the value of the modulus [44–53]. Hence, Young's modulus is essentially a relaxational parameter.

It is postulated herewith that an analytical relation between the creep rate and the relaxation of the Young's modulus may be obtained from Eq. (2). In fact, if Eq. (2) is rewritten for stress, then one gets an expression for the stress relaxation, i.e. its dependence on temperature and deformation rate $\dot{\varepsilon}_p$:

$$\sigma = \frac{U_{0p}}{\gamma_p} - \frac{kT}{\gamma_p}\ln\frac{\dot{\varepsilon}_{0p}}{\dot{\varepsilon}_p}. \tag{4}$$

Dividing by a constant deformation, ε_0, Eq. (4) becomes

$$E = \frac{U_{0p}}{\gamma_p\varepsilon_0} - \frac{kT}{\gamma_p\varepsilon_0}\ln\frac{\dot{\varepsilon}_{0p}}{\dot{\varepsilon}_p} = \frac{U_{0p}}{\gamma_p\varepsilon_0}\left(1 - \frac{kT}{U_{0p}}\ln\frac{\dot{\varepsilon}_{0p}}{\dot{\varepsilon}_p}\right) \tag{5}$$

where $E = \sigma/\varepsilon_0$.

Equation (5) predicts a decrease in Young's modulus with increasing temperature and decreasing deformation rate.

The experiments show that the value of Young's modulus depends strongly on the method of its measurement. The highest E value is obtained from the propagation of hypersonic waves in polymers; lower E values are established in acoustic measurements; still lower E values are obtained in dynamic measurements; and the lowest E values are measured in tensile deformation. Moreover, in each method the E values increase regularly with increasing deformation rate (or frequency) [20].

To generalize these regularities and prove Eq. (5), the following methods and procedures were used to vary $\dot{\varepsilon}_p$: tensile ($\dot{\varepsilon}_p = (10^{-2} - 1)\ s^{-1}$), dynamic ($\dot{\varepsilon}_p = (10^2 - 10^4)$ Hz), acoustic ($\dot{\varepsilon}_p = (10^3 - 10^5)$ Hz), and hyper-sonic ($\dot{\varepsilon}_p \cong 10^{10}$ Hz) tests. For detail, see Refs. [17, 20]. These experiments show that Young's modulus depends on the deformation rate for all oriented polymers.

Fig. 4 shows a typical dependence of Young's modulus vs. $\log \dot{\varepsilon}_p$. At fixed temperatures, the experimental points fall on straight lines which form a "fan". The abscissa of their origin $\dot{\varepsilon}_{0p} \cong 10^{-12}$ Hz. No significant deviations from linearity of the $E(\log \dot{\varepsilon}_p)$ function were observed in the investigated $\dot{\varepsilon}_p$ region. Previously, Assay, Lambertson, and Guenter [54] have shown that there is no dispersion of the modulus in the frequency range ≥ 6 GHz. Therefore, $\nu \cong 6$ GHz may be suggested to be the limiting frequency for realization of the *true* elastic deformation in polymers.

Fig. 5 demonstrates the temperature dependence of Young's modulus for the same polymers as determined by different instrumental methods (different $\dot{\varepsilon}_p$ values). It shows a linear dependence in the temperature range between the main relaxation transitions, regardless of method: line 1 is obtained by an ultrasonic method, line 2 by a dynamic one, and lines 3 and 4 from stress-strain curves. These straight lines also form a "fan", which converges to the E_0 origin at $T \to 0$.

Combination of the dependence presented in Fig. 4 with that shown in Fig. 5 leads to the previously predicted Eq. (5).

Both the present and previously published data [44–53] show that linearity between E and $\ln \dot{\varepsilon}_p$ is a common result for many polymers. Moreover, these results are typical for both compression [46] and shear [53] deformation. Equation (5) is universal, not only for drawn polymers but also for isotropic ones – including rubbers [18, 46, 47].

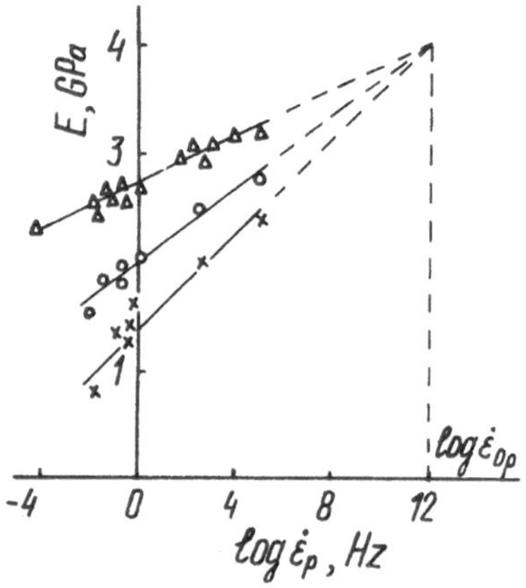

Fig. 4. Young's modulus, E, versus log deformation rate, $\dot{\varepsilon}_p$, for PM S fiber. Temperature (K): (△), 298 (○), 423; and (×), 523

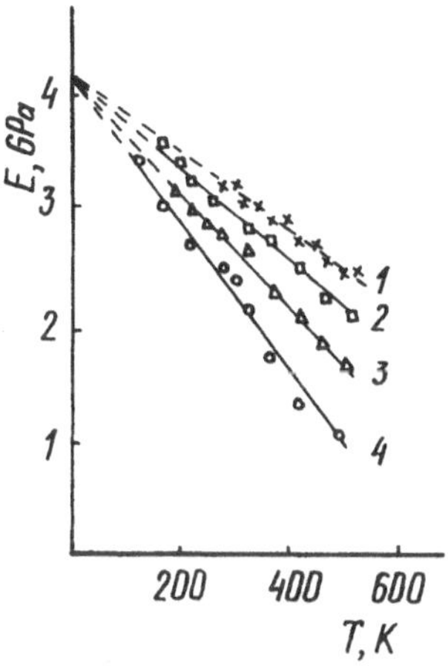

Fig. 5. Young' modulus, E, versus temperature, T, for PM S fiber. Log deformation rate (Hz): 1. (×), 12; 2. (□), 6.4; 3. (△), − 0.1; and 4. (○), − 3.8

To evaluate the limits of its validity, Eq. (5) may be presented in the following form

$$U_{0p} - \gamma_p \sigma = kT \ln \frac{\dot{\varepsilon}_{0p}}{\dot{\varepsilon}_p}. \tag{6}$$

If a stress is too large, $\sigma \cong \sigma_*$, then $\gamma_p \sigma_* \cong U_{0p}$, and Eq. (6) shows no relaxation of Young's modulus, i.e. E = const.

Another limit is governed by the deformation rate: the latter cannot exceed the relative speed of sound c, which is defined as speed of sound divided by sample length (for solid polymers, $c \cong (10^5 - 10^6)\ s^{-1}$). This fact is confirmed by the previously mentioned data: at $\nu \geq 6\ 10^6$ Hz, no dispersion occurs in Young's modulus of polymers [54].

Table 1 presents the activation parameters for the relaxation of Young's modulus.

Finally, relaxation of Young's modulus may be treated as stress relaxation. Naturally, Eq. (6) was shown to be appropriate for stress relaxation. Moreover, its activation parameters are the same as those of the creep rate [24, 55, 56].

3.3 Kinetics of Steady-State Creep

In this Section, the secondary stage of creep, steady-state creep will be considered. In this stage, the creep rate seems to be constant.

Figures 2 and 3 also present the experimental results of the examination of steady-state creep. They are similar to those for primary creep. Therefore, the kinetics of deformation in the secondary creep stage may be expressed by an Arrhenius type equation analogous to that for the primary stage [24]

$$\dot{\varepsilon}_s = \dot{\varepsilon}_{0s} \exp \frac{U_{0s} - \gamma_s \sigma}{kT} \tag{7}$$

where $\dot{\varepsilon}_s$ is the steady-state-creep rate, $\dot{\varepsilon}_{0s} \cong 10^{13}$ Hz, U_{0s} is the activation energy, and γ_s is the activation volume.

Experimental investigations, carried out by Zhurkov [8], and others [9–14], confirm the validity of Eq. (7) for different kinds of polymers and other solids.

The kinetic parameters of Eq. (7) U_{0s} and γ_s, were calculated, and their numerical values are listed in Table 1.

A comparison between Eq. (7), for the kinetics of the steady-state creep, with Eq. (3), for the kinetics of the decelerating creep, leads to the conclusion: the kinetics of creep in both stages may be expressed by a uniform equation. Moreover, the energetic parameters, U_{0s} and U_{0p}, are numerically identical: $U_{0s} \cong U_{0p} = U_0$. Hence, the mechanism of creep in both stages is uniform, and the increase in deformation in both stages is a continuous process [16, 20, 23].

Since the tertiary creep stage, accelerating creep, embraces a short time region, compared with the previous stages, it is difficult to investigate the kinetics of deformation on this stage. The next Section will be devoted to the final phenomenon of creep evolution, the fracture.

3.4 Kinetics of Fracture

As early as 1957, Zhurkov [55] put forward the kinetic concept of the fracture of solids. The main feature of the concept of kinetic fracture is the "durability", τ_f, i.e. the time of loading until fracture. The well known empirical equation of this concept is [8, 9, 55]

$$\tau_f = \tau_{0f} \exp \frac{U_{0f} - \gamma_f \sigma}{kT} \tag{8}$$

where U_{0f} and γ_f are the activation parameters and τ_{0f} is a constant. Investigations of ionic, covalent, and molecular crystals; metals, minerals, and polymers have confirmed its validity in the time range between 10^{-7} and 10^7 s. Equation (8) was established to be suitable for fracture as a result of different types of stress: stretching, compression, tear, and torsion; for cyclic and time-dependent stress; for combined stress (tear, identation), and also for aggressive media and radiation [9, 13].

The pre-exponential factor, τ_{0f}, for all materials is inversely proportional to the period of atomic vibration in solids: $\tau_{0f} = (10^{-12} - 10^{-14})$ s.

Equation (8) may be rewritten for the fracture stress, σ_f, as

$$\sigma_f = \frac{U_{0f}}{\gamma_f}\left(1 - \frac{kT}{U_{0f}} \ln \frac{\tau_f}{\tau_{0f}}\right). \tag{9}$$

It describes fracture stress as a function of temperature and time.

The investigations of the fracture of oriented polymer in the temperature range between the main relaxation transitions show the validity of Eq. (9) [16, 17, 20]. From the experimental dependence, plotted as τ_f versus T^{-1} and as τ_f

versus σ, the kinetic parameters for the fracture rate are calculated and presented in Table 1. When these are compared with the kinetic parameters of steady state creep, they result in: $U_{0f} \cong U_{0s}$ (elsewhere it will be denoted as U_0) and $\gamma_f \cong \gamma_s$ (it will be denoted as γ_0) [24].

Since the duration of the accelerating creep leading to fracture is much less than that of steady state creep, the coincidence between the activation energies and activation volumes of secondary-creep and of fracture is not surprising.

4 Kinetics of Generation and Evolution of Excited Chemical Bonds in Macromolecules as a Background for the Kinetics of Deformation and Fracture

4.1 Kinetics of Deformation and Fracture as a Result of Fluctuations in the Vibrational Energy of Atoms

All kinetic equations presented above have a specific feature: they include an exponential factor exp $[U(\sigma)/\mathbf{k}T] = \exp[U_0 - \gamma\sigma)/\mathbf{k}T]$. As "temperature" is the measure of intensity of atomic vibrations, one must conclude that this motion plays a decisive role in all deformation processes.

A similar Boltzmann factor, $\exp(U/\mathbf{k}T)$, is widely used to describe various physical phenomena. It describes the nonuniform distribution of the energy of atoms in solids. This nonuniformity is caused by the chaotic thermal motion [9].

According to J. Frenkel [5], the average time τ_{fl}, between two subsequent fluctuations of the energetic parameter, U_{fl}, on an atom is equal to

$$\tau_{fl} \cong \tau_0 \exp \frac{U_{fl}}{\mathbf{k}T} \tag{10}$$

The empirical equations displayed above for the kinetics of creep, relaxation of stress and Young's modulus, and of fracture, are assumed to reflect the evolution of energy fluctuations, since they include the Bolzmann factor and are described by Arrhenius type equations. A comparison between the activation parameters of various kinetic processes leads to following conclusions [24]:

(i) All time constants, τ_{0f}^{-1}, $\dot{\varepsilon}_{0p}$, $\dot{\varepsilon}_{0s}$, seem to be similar and equal to $\cong 10^{12}$ Hz.
(ii) The energetic parameters, U_{0p}, U_{0s}, and U_{0f}, are equal for a given polymer.

This is the reason why we conclude that all of the kinetic processes investigated are interrelated and have an uniform origin in energy fluctuations. In other words, energy fluctuations control the kinetics of deformation, relaxation, and fracture.

The regions embracing the excited chemical bonds are certain specific formations, the so called, *dilatons* [30, 32, 33, 39]. In Sect. 4.3, the kinetics of their generation and evolution will be described on the basis of spectroscopic investigations.

4.2 Excited Chemical Bonds in Macromolecules: Their Observation and Investigation

In recent years, excited chemical bonds have been studied experimentally by IR and Raman spectroscopy. To observe the excited chemical bonds, Vettegren [30, 32, 33, 39], Vettegren et al. [24–29, 31, 34–38], and others [56–58] analyzed the shape of the regularity bands in spectra of different polymers. These bands correspond to the vibrations of sequence of chemical bonds in the macromolecule, forming a helix or a planar zigzag. Deformation of valence angles and bonds causes a frequency shift, $\Delta\nu$. The latter is related to the deformation of a helix by the equation [39]:

$$\frac{\Delta\nu}{\nu(0)} = \frac{\nu(\varepsilon) - \nu(0)}{\nu(0)} \cong - G_i\varepsilon \tag{11}$$

where $\nu(\varepsilon)$ and $\nu(0)$ are the frequencies of atomic vibrations in deformed and non-deformed helices respectively, ε is a deformation, and G_i is the Grüneisen parameter. Usually, $G_i > 0$. Therefore, if there are strongly deformed helices in macromolecules, the corresponding elementary bands are expected to appear on the low-frequency wing of the main band. Correspondingly, strongly compressed chemical bonds are expected to appear on a high-frequency wing [26–39].

Indeed, systematic and careful spectroscopic investigations [30, 32, 33, 39] showed that there do exist weak satellites both on the low-frequency wing and (with lower intensity) on the high-frequency wing of the regularity band. They are displayed in Fig. 6. The peaks of the satellites are shifted from the main peak by some tens of cm^{-1} (see Table 2), and they are ca. 1 – 10% as intense as the main band.

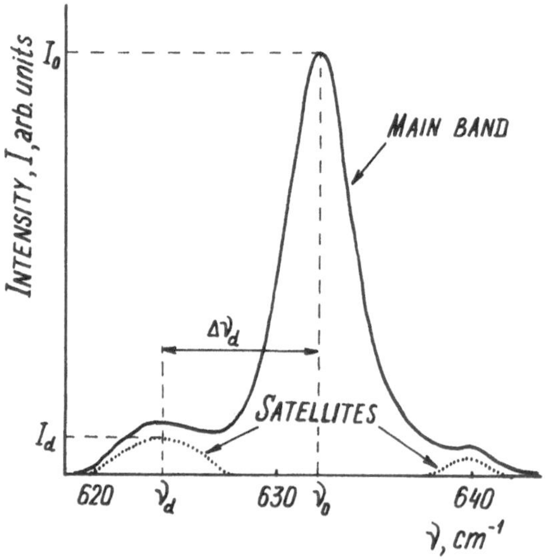

Fig. 6. Regularity band with satellites in the IR spectrum of PETP at 300 K

Table 2. Some characteristics of the regularity bands and deformation of the excited chemical bonds in macromolecules of non-stressed polymers at 300 K

Polymer	Frequency of the regularity band (ν_0, cm^{-1})	Satellite shift, $\Delta\nu_d = \nu_d - \nu_0$ (cm^{-1})	Deformation of the excited bonds (ε_d 10^2)
Nylon 6	930	40	5.5
PP	632	45	4.0
	1614	25	4.2
PE	722	40	6.5
PETP	840	18	4.0
	632	48	4.2
	1614	23	3.8
DPO B	1021	12	4.0
DPO P	1027	17	5.7

We mainly investigated deformation and fracture under tensile stress, which causes an extension of the chemical bonds. For this reason, we mainly examined the low-frequency satellites.

In accordance with Eq. 11, the average extension of the excited chemical bonds, ε_d, is evaluated from the measured frequency shift of the satellite peak, $\Delta\nu_d$, by:

$$\varepsilon_d \cong \frac{\Delta\nu_d}{G_i\nu_d(0)} = -\frac{\nu_d(\varepsilon_d) - \nu_d(0)}{G_i\nu_d(0)} \tag{12}$$

where $\Delta\nu_d = \nu_d(\varepsilon_d) - \nu_d(0)$, is the frequency shift of the satellite.

The relative concentration of the extended excited chemical bonds is determined as a ratio of the intensity of the low-frequency satellite, I_d, and that of the main band, I_0

$$c_d = I_d/I_0. \tag{13}$$

The extension of the excited bonds in the non-stressed sample was discovered to be equal to several per cent, and their concentration is also several per cent of the total concentration of helices in a sample [37, 39].

The spectroscopic methods permit to evaluate the vibrational energy in the regions containing the excited chemical bonds.

In IR spectroscopy, an energy intensity in the emission spectrum, $I_{EM}(\nu)$, is known to be related to that in the absorption spectrum, $I_A(\nu)$, by the equation [61]:

$$I_{EM}(\nu) \cong N I_A(\nu) \tag{14}$$

where N is a phonon occupation number.

In Raman spectroscopy, on the other hand, the energy intensity in the Stokes region, $I_S(\nu)$, is known to be related to that in the anti-Stokes region, $I_{AS}(\nu)$, in

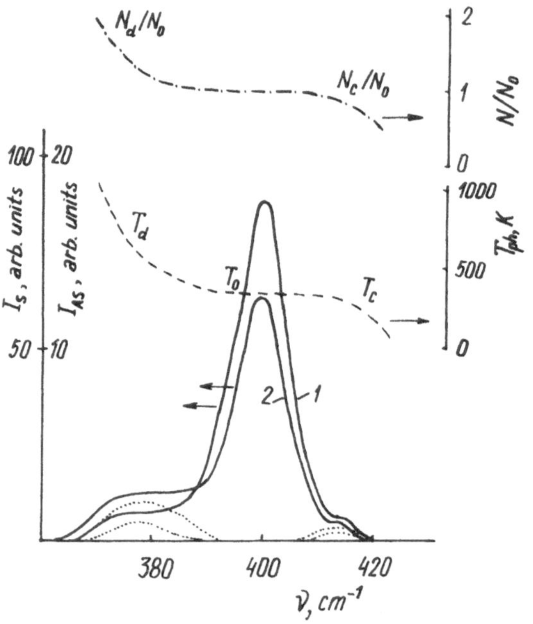

Fig. 7. Regularity band with satellites in the Raman spectrum of PP at 300 K: (1) in the Stokes region and (2) in the anti-Stokes region. The phonon temperature. T_{ph} (– – – –), and the relative phonon occupation number, N/N_0 (–·–·–), are also displayed

the following way [61]:

$$I_S(\nu) \cong \frac{N}{N+1} I_{AS}(\nu) \tag{15}$$

Hence, N can be calculated if $I_{EM}(\nu)$ and $I_A(\nu)$ or $I_S(\nu)$ and $I_{AS}(\nu)$ are known. For details, see Ref. [37].

Figure 7 demonstrates the regularity band of PP in the Raman spectrum and the relative phonon occupation number for the extended excited bonds, N_d/N_0, and that for the compressed excited bonds, N_0/N_0, where N_0 is the equilibrium occupation number. It is clear that $N_d/N_0 > 1$ and $N_c/N_0 < 1$. N is related with temperature by the expression [61]:

$$N \cong \left[\exp\left(\frac{h\nu}{kT_{ph}}\right) - 1\right]^{-1} + \frac{1}{2} \tag{16}$$

The local phonon temperature, T_{ph}, can then be evaluated from Eq. (16) for equilibrium (T_0), extended (T_d), and compressed (T_c) excited bonds. Calculations are shown in Fig. 7: $T_d \cong 1000$ K, $T_c \cong 100$ K, whereas $T_0 = 300$ K. These results prove that the regions containing the stretched dilated bonds are created by energy fluctuations rather than by mechanical stress, because the phonon temperature of the overstressed regions in solids must be equal to the thermodynamic temperature of the sample.

4.3 *Kinetics of Generation and Evolution of the Excited Chemical Bonds in Macromolecules*

Figure 8 shows the evolution in time of ε_d and c_d in strained PP. It is clear that the relaxation of stress (and of Young's modulus) is accompanied by an increase in both the deformation and concentration of the excited chemical bonds.

Figure 9 demonstrates the kinctics of concentration of the excited bonds at both constant stress and constant temperature in the same polymer. As can be seen in Fig. 9, Δc_d increases with time. Moreover, an increase in both T and σ causes an increase in the rate at which the excited bonds accumulate, $\dot{c}_d$. The limiting value of the latter as $t \to 0$, $\dot{c}_d(0)$, can be evaluated as a tangent to the $\Delta c_d(t)$ curve (see Fig. 9). In this case, $\dot{c}_d(0)$ is related to the average time for the generation of the excited bonds, τ_d, in the following way [33, 39]:

$$\tau_d \cong \dot{c}_d(0)/c_0 \tag{17}$$

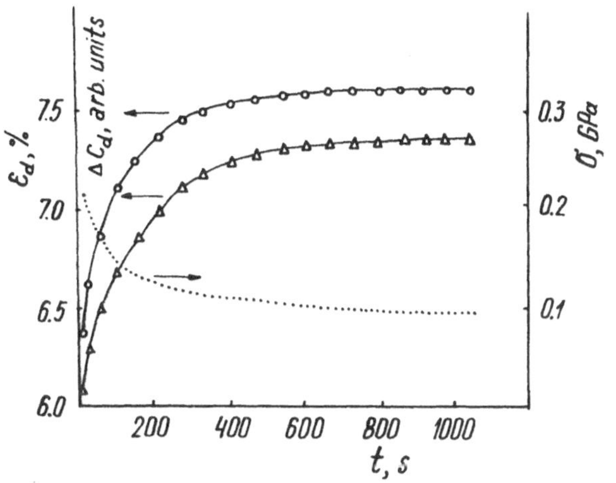

Fig. 8. The increase in concentration, Δc_d ($\triangle$), and deformation, ε_d($\bigcirc$), of the excited chemical bonds versus time, t, as well as stress relaxation (. . . .), for PP at 300 K and $\varepsilon_0 = 2.5\%$

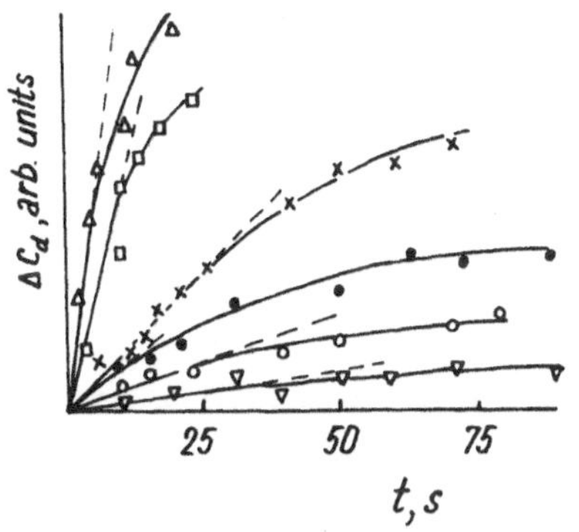

Fig. 9. The increase in the concentration of excited bonds Δc_d, with time, t, evaluated from the IR spectrum of PP ($\triangledown$, $\bigcirc$, $\times$, $\square$, $\triangle$) and BPT B ($\bullet$). Applied stress (GPa)/Temperature (K): ($\triangledown$), 0.45/270; ($\bigcirc$), 0.40/300; ($\times$), 0.45/300; ($\square$), 0.45/320; ($\triangle$), 0.50/300; and ($\bullet$), 0/the increase from 300 to 543

where c_0 is the concentration of helices in a sample. Since c_0 weakly depends on T and σ, the change in $\dot{c}_d$ is supposed to be caused by the dependence of τ_d on both T and σ.

It should be emphasized that the concentration of the excited bonds increases, not only by the combined action of temperature and applied stress, but also in the absence of the latter. Figure 9 shows an increase in $\Delta c(t)$ in the BPTB sample at $\sigma = 0$ with temperature increasing from 300 to 543 K.

The experimental data on the average time for the formation of excited bonds are presented in Fig. 10 as $\log \tau_d$ versus σ at the constant temperatures. The experimental points fall on straight lines which form a "fan". Its origin is at the ordinate value of $\log \tau_{0d} \cong -12$. Figure 11 shows the experimental lines also form a "fan" originating from the same ordinate. Consequently, the average time may be written as [33, 39]

$$\tau_d = \tau_{0d} \exp \frac{U_{0d} - \gamma_d \sigma}{kT} \tag{18}$$

where U_{0d} is the activation barrier to be overcome in order to cause the generation of excited bonds at $\sigma = 0$, and γ_d is the activation volume for the generation of excited bonds. The experimentally calculated activation parameters, U_{0d} and γ_d, are listed in Table 1.

The experiments carried out for the non-stressed polyimide fibers (at $\sigma = 0$) shows that Eq. (18) is valid too. Therefore, excited chemical bonds are generated in the absence of stress.

Figure 12 shows that the dilatation of the excited bonds, ε_d, is observed to increase in proportion to log t, where t is the time since a loading or a temperature increase [37, 39]. The extrapolation of the dependence to $\varepsilon_d = 0$ gives $\log t_{0d} \cong -13$. On the other hand, Fig. 13 shows a proportional increase in ε_d versus temperature [37, 39].

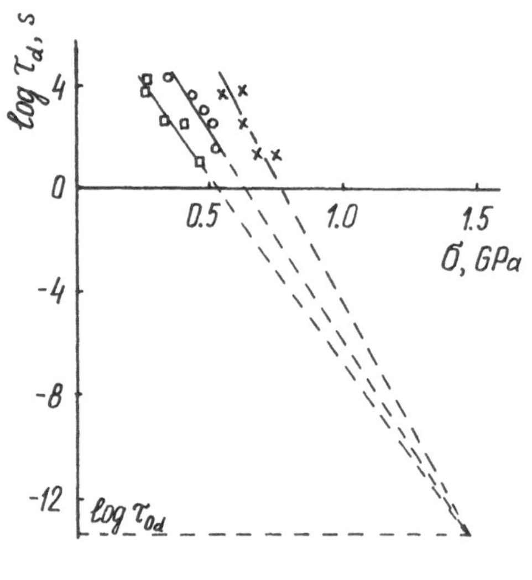

Fig. 10. Log average time for generation of the excited bonds τ_d, versus applied stress, σ, for PP. Temperature (K): (×), 270; (○), 300; (□), 320

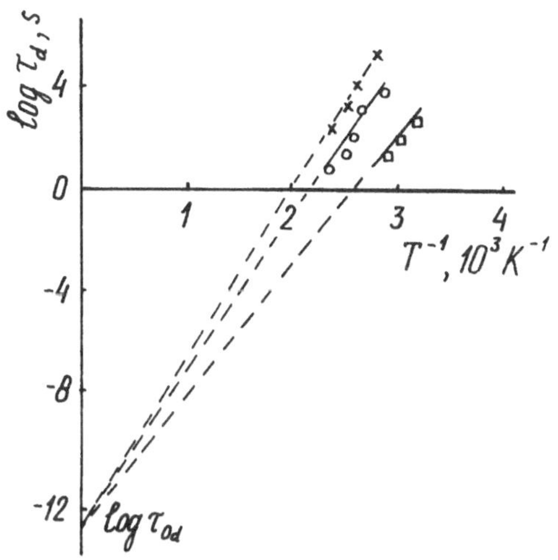

Fig. 11. Log average time for generation of the excited bonds τ_d, versus inverse temperature, T^{-1}, for PP. Applied stress (GPa): (×), 0.40; (○), 0.45; and (□), 0.50

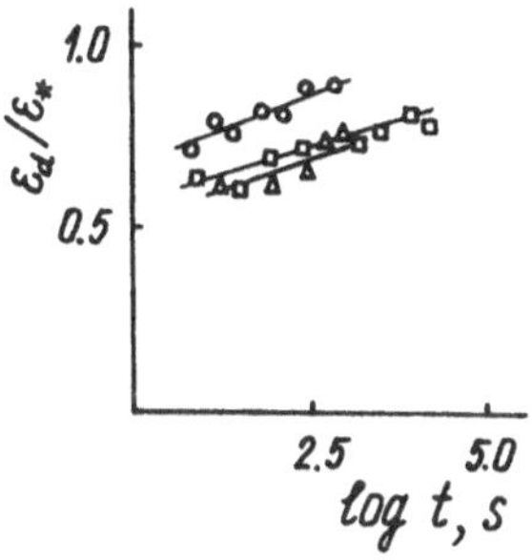

Fig. 12. Deformation of the excited bonds divided by its ultimate value, $\varepsilon_d/\varepsilon_*$, versus log time, t. Polymers: (○, △), PP; and (□), Nylon 6. Applied stress (GPa)/Temperature (K): (△), 0.4/150; (○), 0.6/150; and (□), 0/the increase from 300 to 420

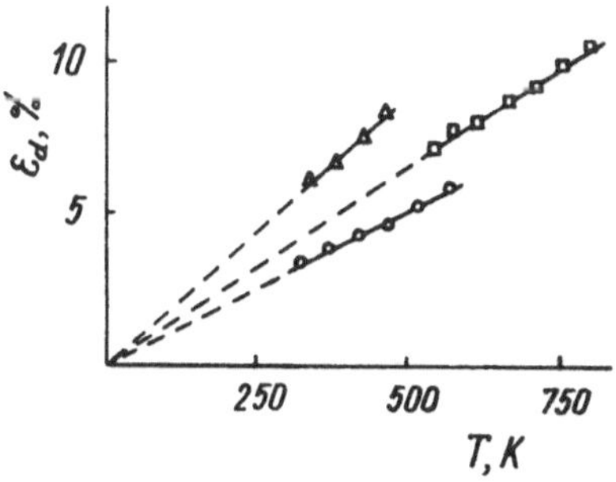

Fig. 13. Deformation of the excited bonds ε_d, versus temperature, T, for non-stressed polymers in 20 min after the temperature set. Polymers: (○), DPO B; (□), DPO P; and (△), PE

The quoted results yield an empirical equation for the description of the extension of excited bonds as a function of time and temperature, after deflection from the equilibrium state [33, 37, 39]

$$\frac{\varepsilon_d}{\varepsilon_*} = \frac{kT\log(t/\tau_{0d})}{U_{0d} - \gamma_d\sigma} \tag{19}$$

where ε_* is the limiting (rupture) deformation of the excited bonds. For most polymers, $\varepsilon_* \cong 0.1$ [33, 37, 39].

As can be seen from Table 1, the activation energy for the generation of excited chemical bonds U_{0d}, coincides with the activation energy U_0 for any macroscopic deformation process, $U_{0d} \cong U_0$. This equality is the principal conclusion.

In the authors' opinion, both deformation and fracture are controlled by the relaxation of some "elementary nuclei". For linear polymers, the elementary nuclei of deformation are supposed to be the gauche conformers [60–65] or the regions with weakened intermolecular bonds. In the latter case, slippage of the macromolecules occurs [66–68]. The nuclei of potential fracture seem to be the "overstressed" regions of macromolecules, in which the valence angles and bonds are stretched, as compared with equilibrium positions.

4.4 Correlation Between Kinetics of Generation of Excited Chemical Bonds with Kinetics of Deformation and Fracture

A careful spectroscopic investigation of the optical intensity, I_{ab}, at the different points of the low-frequency satellite, as a function of time are presented in Fig. 14. They show: (i) at the satellite maximum, $I_{ab} = I_d$ increases with time (curve 1), indicating that, the concentration of excited bonds is increasing too; and (ii) in the low-frequency wing of the satellite, I_{ab} increases only for a certain time, t_h, and then decreases (curves 3 and 4) [27, 64]. Hence, there is a frequency, ν_h, that divides the satellite into two regions: at $\nu < \nu_h$, I_{ab} increases, and at $\nu > \nu_h$, I_{ab} decreases.

To clarify these peculiarities in the satellite shape, one should remember that the shape of a satellite reflects the variation in the elongation of the excited bonds.

Hence, the ν_h parameter devides all extended bonds into two groups. In the first group, $\varepsilon_d < \varepsilon_h = \Delta\nu_h/\nu(0)G_i$, so their concentration increases with time. In the second group, $\varepsilon_d > \varepsilon_h$, and their concentration changes in a more complicated manner: it increases, then achieves a maximum, and finally decreases.

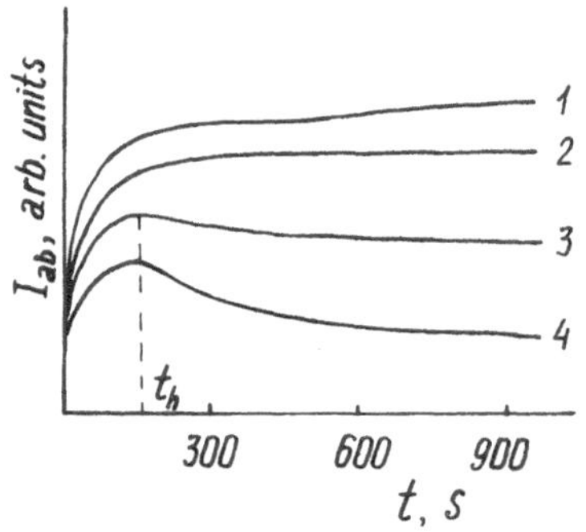

Fig. 14. Absorbance intensity, I_{ab}, in the different points of the low-frequency satellite of the main band $\nu_0 = 975\ cm^{-1}$ in the IR spectrum of PP vs. time, t. Applied stress: 0.42 GPa. $\Delta\nu_d$ $(cm^{-1})/(\varepsilon_d/\varepsilon_*)$: 16/0.4; (2), 20/0.5; (3), 25/0.6; and (4), 30/0.75.

To elucidate the difference between variously elongated excited bonds, their local temperatures, T_d, were evaluated. Figure 15 proves that the phonon temperature of excited bonds depends on ε_d: an increase in ε_d causes an increase in T_d; at $\varepsilon_d \cong \varepsilon_*$, $T_d \cong 2000$ K. Perhaps, these "strongly heated" bonds are the first of all to be dissociated.

To check this supposition, the decrease in concentration of the most stretching bonds, $-\Delta c_d$, was calculated, using the expression:

$$-\Delta c_d \cong \frac{\Delta I_d}{I_d(0)} = \frac{I_d(t) - I_d(t_h)}{I_d(0)}. \tag{20}$$

In Eq. (20), $\Delta I_d = (I_d(t) - I_d(t_h))$ is the decrease in intensity with time and $I_d(0)$ is the satellite intensity at $t = 0$.

Simultaneously, an increase in concentration of new ends of the macromolecules, Δc_e, created by their rupture on account of stress, was investigated by IR spectroscopy [27]. As can be seen from Fig. 16, $-\Delta c_d \cong \Delta c_e$. Therefore, the decrease, $-\Delta c_d$, in concentration of the excited bonds is caused by their dissociation.

The present data indicate that the extended dilated chemical bonds are the basis of the fracture process: the most elongated bonds dissociate and cause

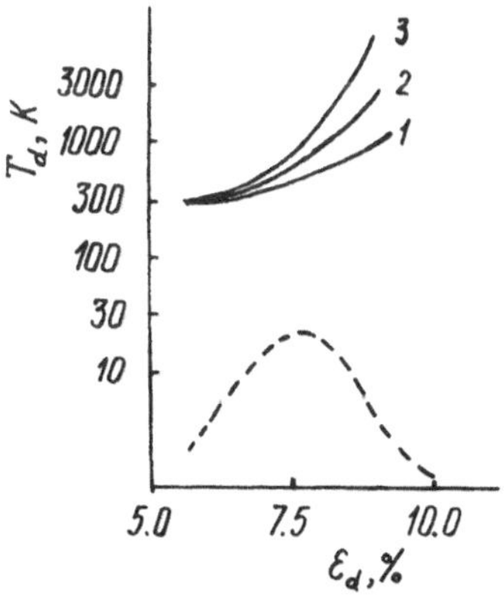

Fig. 15. Phonon temperature, T_{ph}, vs. deformation of the excited bonds, ε_d, for PP at 300 K. Applied stress (GPa): (1), 0; (2), 0.4; and (3), 0.55. The dispersion of the excited bonds (– – –) is also shown

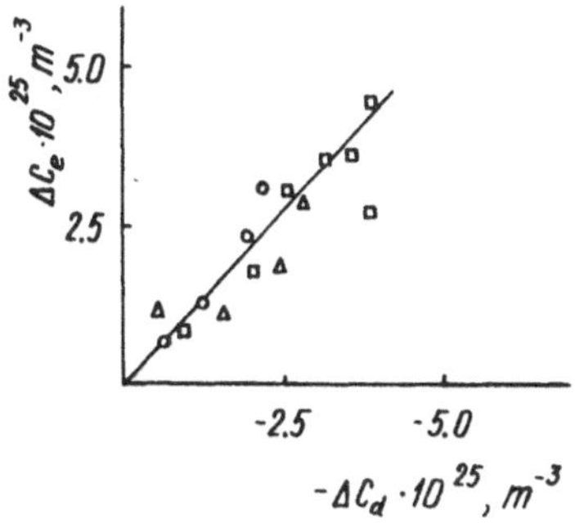

Fig. 16. A comparison between the increase in concentration of new ends of macromolecules, Δc_e, and the decrease in concentration of the excited chemical bonds, $-\Delta c_d$, for PP. Temperature (K)/Applied stress (GPa): (○), 150/0.5; (□), 400/0.3; and (△), 370/0.475

"elementary fracture nuclei". Hence, the expectation time for the generation of excited bonds, τ_d, is correlated with the time to fracture, τ_f.

To check this conclusion, a comparison between τ_d and τ_f was made for the PP samples tested under identical conditions: $\sigma = 0.45$ GPa and elevating the temperature from 270 to 320 K. Figure 17 confirms the equality between τ_d and τ_f.

Remember that the activation energy of fracture is equal to that of creep. Therefore, the molecular mechanism causing both processes may well have the same origin. Indeed, a comparison between the kinetic parameters for the generation of excited bonds, U_{0d} and γ_d, and those of creep, U_{0s} and γ_s, listed in Table 1 for perfectly drawn polymers, shows: $U_{0d} \cong U_{0s}$ and $\gamma_d \cong \gamma_s$, as well as $\tau_{0d} \cong \dot{\varepsilon}_{0s}^{-1}$. Therefore, the average time for the generation of excited bonds controls the kinetics of deformation and fracture.

To correlate the kinetics of macroscopical processes (fracture and relaxation of Young's modulus) with the kinetics of microscopical processes (extension of the excited chemical bonds), let us substitute Eq. (19) into Eqs. (5) and (9). If the equalities $U_{0p} = U_{0d}$ and $\tau_{0p} = \tau_{0d}$ are supposed to be valid, then Eqs. (5) and (9) become:

$$E = E_0\left(1 - \frac{\varepsilon_d}{\varepsilon_*}\right) \tag{21}$$

and

$$\sigma_f = \sigma_{0f}\left(1 - \frac{\varepsilon_d}{\varepsilon_*}\right) \tag{22}$$

Equations (21) and (22) predict linearity between E (or σ_f) and the deformation of the excited bonds.

In Fig. 18, E and σ_f are compared with ε_d for oriented polyimide fibers in the temperature range from 100 to 750 K. They indeed demonstrate linear dependence as predicted by Eqs. (21) and (22). This result is supposed to prove a correlation between macroscopic processes with the elongation of the excited bonds in macromolecules.

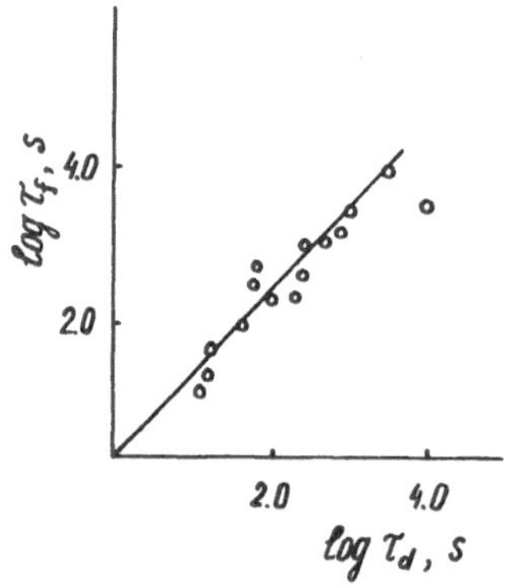

Fig. 17. Log time to fracture, τ_f, versus log average time for generation of the excited bonds, τ_d, for PP at $\sigma = 0.45$GPa and temperature varying from 270 to 320 K

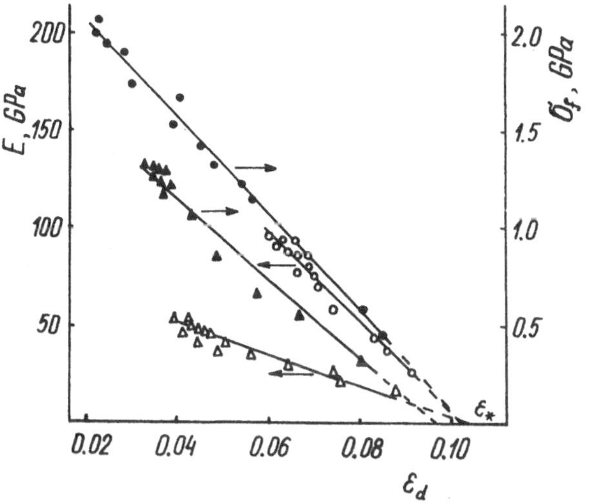

Fig. 18. Young's modulus, E (○, △), and tensile strength, σ_f (●, ▲), as compared with the excited-bond deformation, ε_d, in the temperature range from 100 to 750 K. Polymers: (○, ●), DPO B; and (△, ▲), DPT B

The slopes of these functions are $dE/d\varepsilon_d = -E_0/\varepsilon_*$ and $d\sigma/d\varepsilon_d = -\sigma_{0f}/\varepsilon_*$. The ε_* parameter estimated from these slopes yields $\varepsilon_* \cong 0.1$, and their extrapolation to $E = 0$ and $\sigma_f = 0$ results in the same value: $\varepsilon_* \cong 0.1$. Hence, the limiting deformation of the excited bonds appears to be $\cong$ 10%-as is generally supposed.

We also wish to postulate that the rates of both deformation and fracture are caused by the rate of excited-bond generation rather than the rate of their dissociation. Therefore, the activation energy, U_0, characterizes the generation of the excited chemical bonds rather than their relaxation.

Perhaps, the generation of the excited extended bonds in any part of a macromolecule causes a compression and a partial twisting of its neighbouring parts. Indeed, for a regularity band, a low-frequency satellite corresponding to the excited elongated bonds is partially "balanced" by a high-frequency satellite related to the compressed bonds. Both elongation and compression are caused by a change in valence angles and bonds in helices. Besides, there is another way to decrease a molecule's length: through conformational *trans-gauche* transitions. To accomplish this, the valence angles in the neighbouring parts (consisting of *trans* isomers) must be temporarily increased.

This conclusion is not quite original. As early as the 1930s, Kuhn [71] and Fuoss and Kirkwood [72] suggested that conformational transitions in certain local regions of a macromolecule should be accompanied by deformation of valence angles and bonds in other regions. Similar theoretical models of polymer deformation were considered by Schatzki [73], Boyer [74], Pechhold [75, 76] Gotlib et al. [77–79], and Robertson [80].

5 Kinetics of Deformation, Relaxation, and Fracture in Polymers over a Wide Temperature Range

5.1 The Problems to be Solved

In Sect. 3, the kinetics of deformation, relaxation, and fracture was investigated only between two specific temperatures of the main relaxation transitions. In Sect. 4, the kinetics of these processes, described by Eqs. (3), (5), (7), and (8) was shown to be strongly related to the generation of excited chemical bonds, caused in turn by energy fluctuations. Nevertheless, the experimental results in a wide temperature range show that, in general, Eq. (1) is not obeyed.

For instance, Fig. 19 presents the temperature dependence of Young's modulus and of tensile strength in a wide temperature range. They will be considered more thoroughly in Sect. 5.3. Nevertheless, a rough examination testifies that at low temperatures, $0 < T < T_t$, the E(T) function is not linear. Hence, it cannot be described by Eq. (5) if U_0 and γ are assumed to be constant. Eq. (5) is probably fulfilled in the $T_t < T < T_b$ and $T_b < T < T_{melt}$ regions, where E(T) is a linear function. However, its slope in the regions mentioned above is different: $(\partial E/\partial T)_t < (\partial E/\partial T)_b$. Therefore, a distinction between the activation parameters for both regions should be made.

Indeed, Eq. (5) and Fig. 19 present the following expressions for the kinetic parameters in both regions. In the $T_t < T < T_b$ region, the activation energy,

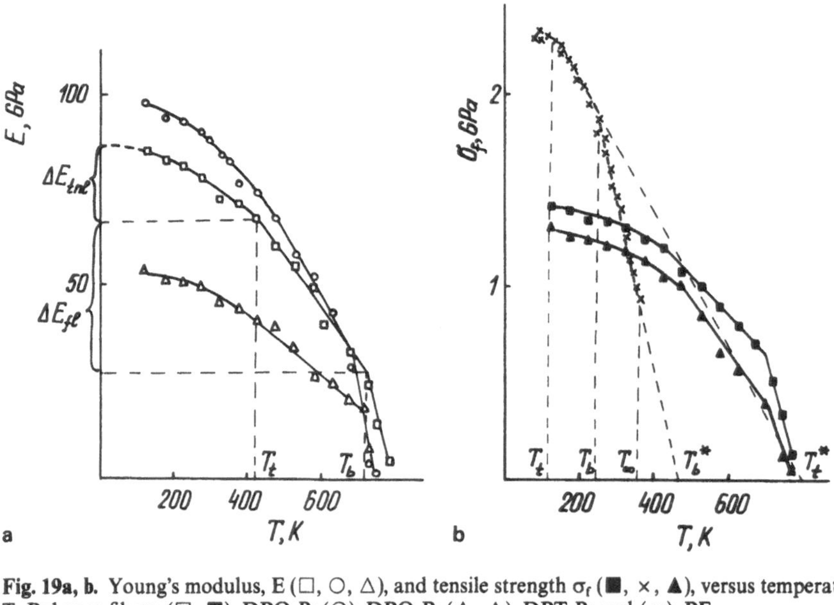

Fig. 19a, b. Young's modulus, E (□, ○, △), and tensile strength σ_f (■, ×, ▲), versus temperature, T. Polymer fibers: (□, ■), DPO B; (○), DPO P; (△, ▲), DPT B; and (×), PE

U_{0t}, and the activation volume, γ_t, are calculated in the following way:

$$U_{0t} = kT_t^* \ln \frac{\dot{\varepsilon}_0}{\dot{\varepsilon}} \tag{23}$$

and

$$\gamma_t = \frac{k}{(\partial E/\partial T)_t} \ln \frac{\dot{\varepsilon}_0}{\dot{\varepsilon}} \tag{24}$$

In the $T_b < T < T_{melt}$ region, they can be written:

$$U_{0b} = kT_b^* \ln \frac{\dot{\varepsilon}_0}{\dot{\varepsilon}} \tag{25}$$

and

$$\gamma_b = \frac{k}{(\partial E/\partial T)_b} \ln \frac{\dot{\varepsilon}_0}{\dot{\varepsilon}} \tag{26}$$

In Eqs. (23)–(26), T_t^* and T_b^* are the extrapolated temperatures to $E = 0$ from the $T_t < T < T_b$ and $T_b < T < T_{melt}$ regions, respectively; $(\partial E/\partial T)_t$ and $(\partial E/\partial T)_b$ are the slopes of the E(T) dependence in the corresponding regions. Calculations using Eqs. (23)–(26) are plotted in Fig. 20. It is clear that $U_{0t} > U_{0b}$ and $\gamma_t > \gamma_b$.

Another way to calculate the activation parameters of creep is Dorn's method [6, 7, 81, 82]. If $\dot{\varepsilon}_1$ and $\dot{\varepsilon}_2$ are the creep rates corresponding to either the

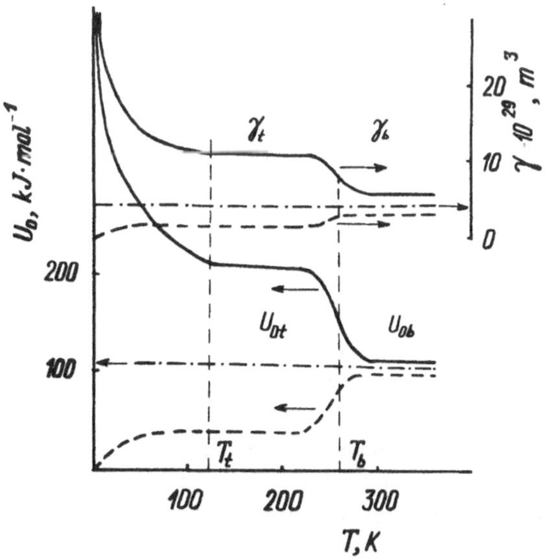

Fig. 20. Activation energy, U_0, and activation volume, γ_0, at different temperatures, T, for PE. Calculations by Eqs. (23)–(26) (———) and by Eqs. (27) and (28) (– – –). The true activation parameters calculated by Eqs. (58)–(60) are also displayed (–·–·–)

"stress jump" from σ_1 to σ_2 (at T = const) or the "temperature jump" from T_1 to T_2 (at σ = const), then Eq. (7) gives the following expressions for the activation parameters

$$U_0 = \mathbf{k}T^2 \left.\frac{\ln(\dot{\varepsilon}_2/\dot{\varepsilon}_1)}{T_2^{-1} - T_1^{-1}}\right|_\sigma + \gamma\sigma \tag{27}$$

and

$$\gamma = \mathbf{k}T \left.\frac{\ln(\dot{\varepsilon}_2/\dot{\varepsilon}_1)}{\sigma_2 - \sigma_1}\right|_T \tag{28}$$

The activation parameters calculated by Eqs. (27) and (28) are presented in Fig. 20 as the dashed plots. It is obvious that $U_{0t} < U_{0b}$ and $\gamma_t < \gamma_b$.

Fig. 20 clears up a difference between the parameters calculated from E(T) and $\dot{\varepsilon}(T)$ in the $0 < T < T_b$ region. Naturally, the former are greater and the latter are smaller than those in the $0 < T < T_b$ region. Hence, there are two problems to be solved.

(i) It is impossible to describe the kinetics of deformation in a wide temperature range by an Arrhenius-type equation with constant activation parameters.
(ii) Different methods of calculation result in distinctive values of the activation parameters.

In this Section, an approach to the description of kinetics of deformation, relaxation, and fracture of polymers over a wide temperature range will be elaborated on the basis of kinetic equations with constant activation parameters. A nonuniform distribution of the energy of atoms over the degrees of freedom requires the replacement of T by the $F(\theta/T)$ quantum function. The resultant (modified) Arrhenius equation describes adequately both deformation and fracture of polymers in the range from $\cong$ 20 K to the melting temperature.

5.2 The Quantum $F(\theta/T)$ Function and Thermal Properties of Macromolecules in the Condensed State

The basic Eq. (1) for physical kinetics implies a uniform distribution of the vibrational energy of atoms over all degrees of freedom, i.e. the average energy of a degree of freedom is **k**T. Nevertheless, for most polymers, this statement is not fulfilled in the region below the melting temperature, T_{melt}.

Indeed, the maximum frequency of atomic vibrations in macromolecules, as determined from IR, Raman, and neutron inelastic scattering spectra reaches 2800–3700 cm^{-1} [83–85]. To excite a vibration of such a frequency, an energy $W = h\nu$ is needed, i.e. a polymer has to be heated up to the thermodynamic Debye temperature $\theta = h\nu/\mathbf{k}$ = 4000–5000 K. At $T \geq \theta$, the energy is distributed uniformly over all fundamental vibrations according to Maxwell-Boltzmann statistics, i.e. each vibrational mode has the energy $W = \mathbf{k}T$. At $T < \theta$, the energy is known to be distributed over vibrations according to Bose-

Einstein statistics [86]

$$W = \frac{h\nu}{\exp\left(\frac{h\nu}{kT} - 1\right)} + \frac{h\nu}{2}. \tag{29}$$

For macromolecules of regular structure, *m* degrees of freedom are equivalent to *m* nodes of fundamental vibrations. Each *m*-th vibrational mode includes *n* fundamental vibrations; *n* is equal to the number of crystal cells in a macromolecule. The wave length of the vibrations varies from the double length of the molecule to the double length of a crystal cell, 2a. At the same time, the frequency varies from $\nu_{mn}^{min} \cong 0$ to $\nu_{mn}^{max} \cong c_f/2a$ where c_f is the wave velocity. The energy of the *m*-th vibrational mode is the sum of the energies of all vibrations of this mode:

$$W_m = \sum_n \left[\frac{h\nu_{mn}}{\exp\left(\frac{h\nu_n}{kT}\right) - 1} + \frac{h\nu_{mn}}{2} \right] \tag{30}$$

At any temperature $T < \theta$, all vibrations in solids may be divided into two groups.

(i) The first group contains vibrations for which the mode Debye temperature $\theta_m \leq T$. The mode Debye temperature, θ_m, is determined from the maximum frequency of the *m*-th mode, ν_{mn}^{max}

$$\theta_m = h\nu_{mn}^{max}/\mathbf{k} \tag{31}$$

For these modes, $\exp(h\nu_{mn}/\mathbf{k}T) \cong 1 + h\nu_{mn}/\mathbf{k}T$, and their energy $W_m \cong \mathbf{k}T$ is the same for all *m* vibrations.

(ii) The second group contains the vibrations for which $\theta_m > T$, i.e. the energy of the *m*-th mode, $W_m < \mathbf{k}T$.

To take into account the non-uniform energy distribution over the degrees of freedom, Eq. (1) should be modified. For this purpose, Gilman [87] and Salganik [88] have introduced the quantum function $F(\theta/T)$ instead of T. Then Eq. (1) is rewritten as

$$k = k_0 \left[-\frac{U(\sigma)}{\mathbf{k}F(\theta/T)} \right] \tag{32}$$

The frequencies of the fundamental vibrations are known to depend on both the chemical and conformational structures of the macromolecule. Therefore, there is no general analytical expression for the $F(\theta/T)$ function.

For approximation calculations, Slutsker et al. [89, 90] have modelled a polymer as an atomic chain. In this case, $F(\theta/T)$ may be treated as series of more than 10 terms.

There is another way to calculate the $F(\theta/T)$ function. For this purpose, we used the phenomenon of the frequency shift with temperature of the fundamental vibrations in IR and Raman spectra of polymers.

For crystals, the $F(\theta/T)$ function was shown [91] to be related to the linear expansion, ε, and the thermal expansion coefficient at $T > \theta$, e, by a relationship

$$F(\theta/T) = \frac{\varepsilon(T)}{e}. \tag{33}$$

As to polymers, their thermal expansion is caused by: (i) a change in the distance between macromolecules, (ii) a change in the conformation of the macromolecules, (iii) an increase in the hole concentration, and (iv) extension of the chemical bonds. It is too difficult to evaluate precisely the contribution of each increment in the thermal expansion. Therefore, to calculate the thermal expansion of a polymer, Vettegren et al. suggested a special spectroscopic procedure. They measured the axis length of helices in a macromolecule. A change in the axis length is caused by an increase in valence angles and bond length under heating. This leads to a band shift in IR and Raman spectra of polymers [39, 92, 93].

For this method, the most suitable bands are those that correspond to the skeletal vibrations of helices in macromolecules. Besides, they should be insensitive to the change in intermolecular forces with increasing temperature. For such vibrations, the band shift is connected to a relative change in the axis length, ε_T, by the relationship:

$$\Delta\nu(T) = \nu(T) - \nu(T \to 0) = -G_1\nu(T \to 0)\varepsilon_T \tag{34}$$

In Eq. (34), $\nu(T \to 0)$ is the extrapolated ν value at $T \to 0$. It results in the expression:

$$\varepsilon_T = -\frac{\Delta\nu(T)}{G_i\nu(T \to 0)} \tag{35}$$

Substitution of Eq. (35) into Eq. (33) yields an expression for $F(\theta/T)$ as a function of $\Delta\nu(T)$:

$$F(\theta/T) = \frac{\Delta\nu(T)}{(\partial\nu/\partial T)\nu(T \to 0)} \tag{36}$$

where $(\partial\nu/\partial T)$ is the temperature coefficient of the frequency shift in the $T > \theta$ region.

At $T > \theta$, the thermal expansion coefficient, e, as known from thermodynamics, is independent of temperature. Therefore, $(\partial\nu/\partial T) = -G_ie = \text{const}$. Consequently, the frequency shift, $\Delta\nu(T)$, and the $F(\theta/T)$ function are directly proportional to one another.

$$F(\theta/T) = \text{const}\,\Delta\nu(T). \tag{37}$$

In Figs. 21 and 22, the regularity-band frequencies, ν, and the $F(\theta/T)$ function are plotted as function of temperature.

At $T \to 0$, $F(\theta/T)$ tends to a constant, $F(0)$; at $T_t < T < T_b$, it changes linearly; and at $T_b < T < T_{melt}$, it is also linear, but with a larger proportionality coefficient.

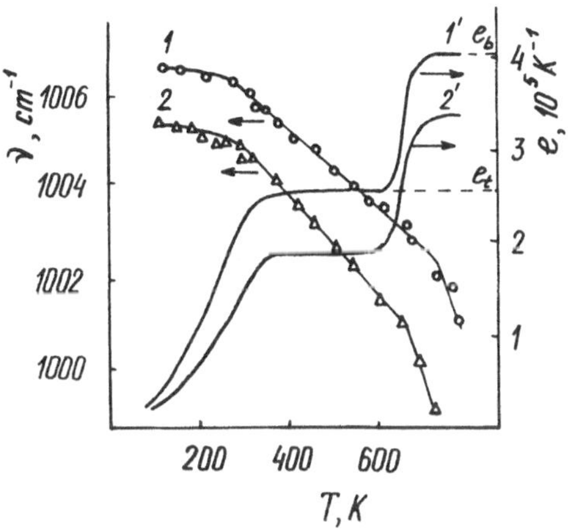

Fig. 21. Regularity-band frequency, ν (1, 2), and coefficient of thermal expansion, e(1′, 2′), versus temperature, T. Polymers: (1, 1′), DPO B; and (2, 2′), DPT B

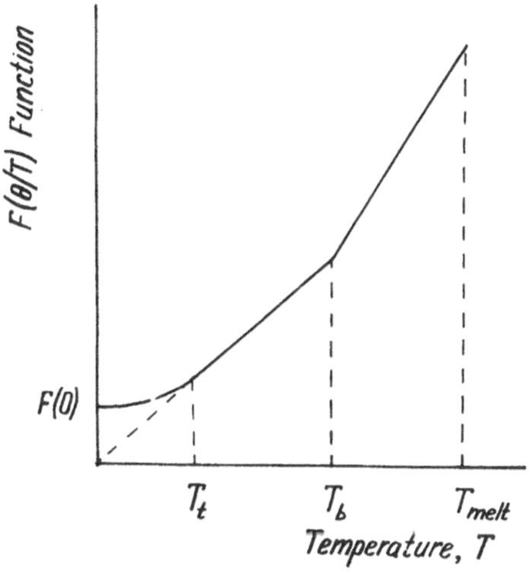

Fig. 22. The F(θ/T) function versus temperature, T

Such a complicated character of F(θ/T) is clarified in the theory of heat capacity for linear polymers, established by Dole [94] and described in detail by Wunderlich and Baur [95]. It is based on a concept advanced by Tarasov [96] and developed by Lifshitz [97] for chain solids.

The thermal expansion, ε_T, caused by atomic vibrations is known to be expressed as the partial derivative of the free energy F with respect to stress [97]:

$$\varepsilon_T = -\frac{\partial F}{\partial \sigma} \tag{38}$$

In the quasi-harmonic theory, F is a function of temperature [98]:

$$F = U_L + kT \sum_{m,n} \ln\left[1 - \exp\left(-\frac{h\nu_{mn}}{kT}\right)\right] + \sum_{m,n} \frac{h\nu_{mn}}{2} \tag{39}$$

where U_L is the energy of atomic interaction. Then, differentiation of free energy results in an expression for ε_T:

$$\varepsilon_T = -\frac{S}{V}\left\{\sum_{m,n} G_{mn}\frac{h\nu_{mn}}{2} + \sum_{m,n} G_{mn}\frac{h\nu_{mn}}{\left[\exp\left(\frac{h\nu_{mn}}{kT}\right)-1\right]^2}\right\} \tag{40}$$

where S is the isothermal compliance, V is the cell volume, and $G_{mn} = -\partial(\ln \nu_{mn})/\partial(\ln V)$ is the calculations, the Grüneisen parameters for all vibrations of the m-th mode are assumed to be equal, i.e. $G_{mn} = G_m$. Besides,

$$\sum_n h\nu_{mn} \cong \frac{k\theta_m}{2}. \tag{41}$$

Then Eq. (40) becomes:

$$\varepsilon_T = -\frac{S}{V}\left\{\sum_m G_m\frac{k\theta_m}{4} + \sum_{m,n} G_m\frac{h\nu_{mn}}{\left[\exp\left(\frac{h\nu_{mn}}{kT}\right)-1\right]^2}\right\} \tag{42}$$

Differentiation of Eq. (42) with respect to temperature gives an equation for the thermal expansion coefficient

$$e(T) = -\frac{S}{V}\sum_m G_m C_m(T) \tag{43}$$

where $C_m(T)$ is the contribution of the m-th mode to the heat capacity described as [95, 98]

$$C_m(T) = \mathbf{k}\sum_n\left(\frac{h\nu_{mn}}{kT}\right)^2\frac{\exp(h\nu_{mn}/\mathbf{k}T)}{\left[\exp\left(\frac{h\nu_{mn}}{kT}\right)-1\right]^2} \tag{44}$$

To calculate $C_m(T)$, its ultimate value should be taken into account: (i) at $T \to 0$, $C_m \to 0$; and (ii) at $T \geq \theta_m$, $C_m = \mathbf{k}$. However, at $T \cong \theta_m/3$, C_m almost attains the classic value: $C_m = (0.7 - 0.8)\mathbf{k}$ [95, 98]. Hence, already at $T \cong \theta_m/3$, $C_m \cong \mathbf{k}$.

Combination of Eqs. (42) and (44) results in the expression for $F(\theta/T)$:

$$F(\theta/T) = \frac{1}{GC}\left[\sum_m G_m\frac{k\theta_m}{4} + \sum_m G_m\int_0^T C_m(T)\,dT\right] \tag{45}$$

In Eq. (45), G is the thermodynamic Grüneisen parameter, $G = \Sigma G_m/\Sigma m$, and C is the "classic" heat capacity at $T > \theta$, i.e. $C = i\mathbf{k}$.

Figure 22 shows that $F(\theta/T)$ is a linear function within two regions: at $T_t < T < T_b$ and at $T_b < T < T_{melt}$. Hence, only two modes (with the ν_t and ν_b maximum frequencies) give the greatest contribution to the $F(\theta/T)$ function.

The maximum frequencies, ν_t and ν_b, can be calculated from either the E(T), $\sigma_f(T)$, or $\nu(T)$ dependence. The characteristic T_t and T_b temperatures determined from these dependences form the basis for the ν_t and ν_b calculations. Indeed, $T_t = \theta_t/3 \cong h\nu_t/3\mathbf{k}$ and $T_b = \theta_b/3 \cong h\nu_b/3\mathbf{k}$. Therefore, $\nu_t \cong 3\mathbf{k}T_t/h$ and $\nu_b \cong 3\mathbf{k}T_b/h$ [40].

The calculated ν_t and ν_b parameters are listed in Table 3. They coincide with the maximum frequencies of certain modes of fundamental vibrations as determined from neutron inelastic scattering spectra [83, 84], IR and Raman spectra [85], and theoretical calculations [99, 100].

Table 3 shows that for PE, PP, and Nylon 6, the ν_t values are close to the frequencies of the skeletal twisting vibrations. For aromatic polyesters and polyimides, the ν_t parameters correspond to the vibrations of the benzene ring, in combination with the deformation vibrations of the C–C and C–N bonds.

The ν_b values for PE, PP, and Nylon 6 may be attributed to the maximum frequencies of the skeletal deformation vibrations. For aromatic polyesters and polyimides, they coincide with the vibrations of the benzene ring, in combination with the valence vibrations of the C–C and C–N bonds.

Equation (45) shows that the contribution of any vibrational mode to the $F(\theta/T)$ function is proportional to the mode Grüneisen parameter. Therefore, the mode G_t and G_b parameters for the ν_t and ν_b vibrations are assumed to be the largest in comparison with other mode Grüneisen parameters.

To justify this conclusion, let us turn our attention to the temperature dependence of the thermal-expansion coefficient of the axis length of the helices in a macromolecule, e(T). A e(T) function may be obtained by differentiation of

Table 3. Characteristic temperatures, T_t and T_b, and maximum frequencies of certain vibrational modes, ν_t and ν_b

Polymer	Characteristic temperatures as determined from E(T) and σ_f(T)		Frequencies of certain vibrational modes as determined			
			from E(T) and σ_f(T)		from neutron scattering ([a]) [83, 84] and IR ([b]) spectroscopy	
	T_t(K)	T_b(K)	ν_t (cm^{-1})	ν_b (cm^{-1})	ν_t (cm^{-1})	ν_b (cm^{-1})
Nylon 6	140	390	290	810	340[a]	760[a]
PAN	160	380	330	790	340[a]	800[a]
PE	120	250	250	520	340[a]	520[a]
PEO	260	360	540	750	540[a]	870[a]
PETP	170	390	350	810	280[b]	800[b]
POM	260	400	540	830	470[a]	840[a]
PP	220	280	460	580	470[a]	570[a]
PTFE	160	270	330	560	345[a]	560[a]
BPT B	420	670	870	1390	850[b]	1420[b]
DPO B	420	720	870	1500	840[b]	1460[b]
DPO P	410	650	850	1350	800[b]	1340[b]
DPT B	470	660	980	1370	940[b]	1400[b]
PM	390	650	810	1350	810[b]	1370[b]

a $\nu(T)$ function over temperature in the following way:

$$e(T) = \frac{\partial \varepsilon}{\partial T} = -\frac{\partial \nu}{G_m \nu(T \to 0)\, \partial T} \tag{46}$$

As is readily seen from Fig. 21, at low temperatures ($0 < T < T_t$) e is too small ($e \cong 0$). At $T \cong T_t$, e increases abruptly to the e_t value ($e \cong e_t$). In the $T_t < T < T_b$ region, e is a constant, but at $T \cong T_b$, it increases again to a new constant value ($e \cong e_b$). It remains invariable in the $T_b < T < T_{melt}$ region. Hence Fig. 21 confirms the conclusion that the G_t and G_b mode Grüneisen parameters for the ν_t and ν_b vibrations have the greatest values among the other mode Grüneisen parameters.

5.3 Kinetics of Deformation, Relaxation, and Fracture, Treated with the Quantum Function

Figure 19 demonstrates the temperature dependences of Young's modulus and tensile strength over a wide temperature range. It is clear that they are linear only within the $T_t < T < T_b$ and $T_b < T < T_{melt}$ regions. The T_t and T_b temperatures correspond closely to the β and α relaxation transitions, respectively. In the $0 < T < T_t$ region, deviations from linearity become apparent [34, 35, 101–106].

To describe the kinetics of deformation over a wide temperature range, the $F(\theta/T)$ function should be taken into account instead of T. For instance, Young's modulus is expressed by:

$$E = \frac{U_0}{\gamma\varepsilon_0} - \frac{kF(\theta/T)}{\gamma\varepsilon_0} \ln\frac{\dot{\varepsilon}_0}{\dot{\varepsilon}}. \tag{47}$$

Substitution of Eq. (36) in Eq. (47) yields

$$E = \frac{U_0}{\gamma\varepsilon_0} - \frac{k\nu(0)}{\gamma\varepsilon_0(\partial\nu/\partial T)} \ln\frac{\dot{\varepsilon}_0}{\dot{\varepsilon}} + \frac{k\nu(T)}{\gamma\varepsilon_0(\partial\nu/\partial T)} \ln\frac{\dot{\varepsilon}_0}{\dot{\varepsilon}} \tag{48}$$

Eq. (48) shows that the E(T) function is proportional to the $\nu(T)$ function. Indeed, E(T) and $\nu(T)$ are similar, as can be seen from Figs. 19 and 21. Besides, Fig. 23 demonstrates the proportionality between E and ν in the region from 100 to 750 K. This result proves the proposition that the change in the statistics of fundamental vibrations, described by the $F(\theta/T)$ function, causes deviations from linearity in the E(T) dependence as predicted by Eq. (48). Simultaneously, it provides an opportunity to correlate the kinetics with peculiarities of the different vibrational modes. Substitution of Eq. (45) into Eq. (48) results in

$$E = \frac{U_0}{\gamma\varepsilon_0} - \sum_m \frac{G_m k\theta_m \ln(\dot{\varepsilon}_0/\dot{\varepsilon})}{4G\gamma\varepsilon_0} - \sum_m \frac{kG_m}{G} \int_0^T \frac{C_m(T)dT \ln(\dot{\varepsilon}_0/\dot{\varepsilon})}{C\gamma\varepsilon_0} \tag{49}$$

Equation (49) adequately describes Young's modulus as a function of temperature and loading rate over all investigated temperatures. It provides an oppor-

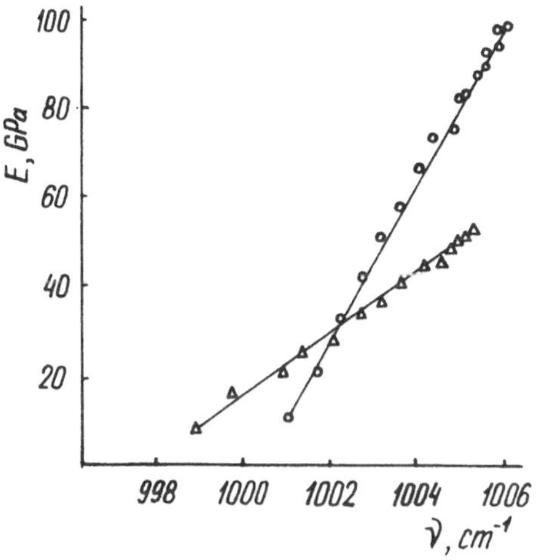

Fig. 23. A comparison between Young's modulus, E, and regularity-band frequency, ν, in the temperature range from 100 to 750 K. Polymers: (○), DPO B; and (△), DPT B

tunity to evaluate the contribution of different vibrational modes to the relaxation of Young's modulus.

A similar equation can be written for the tensile strength:

$$\sigma_f = \frac{U_0}{\gamma} - \sum_m \frac{G_m k\theta_m \ln(\tau_f/\tau_0)}{4G\gamma} - \sum_m \frac{kG_m}{G} \int_0^T \frac{C_m(T)dT \ln(\tau_f/\tau_0)}{C\gamma} \tag{50}$$

Also, writing Eq. (49) for the creep rate, $\dot{\varepsilon}$, results in:

$$\ln\frac{\dot{\varepsilon}_0}{\dot{\varepsilon}} = \frac{U_0 - \gamma\sigma}{\mathbf{k}\sum_m \left[\frac{G_m\theta_m}{4G} + \frac{G_m}{GC}\int_0^T C_m(T)dT\right]} \tag{51}$$

Equations (49)–(51) show that the contribution of the vibrational mode relaxation of Young's modulus, fracture and creep is characterized by the mode Grüneisen parameter and the maximum frequency of the vibrational mode.

Equation (51) shows that at $T \to 0$, the creep rate, $\dot{\varepsilon}(0)$, has a finite value different from zero:

$$\dot{\varepsilon}(0) = \dot{\varepsilon}_0 \exp\left(-\frac{U_0 - \gamma\sigma}{\frac{\mathbf{k}}{4G}\sum_m G_m\theta_m}\right) \tag{52}$$

At the same time, at $T \to 0$, both Young's modulus, E(0), and strength, $\sigma_f(0)$, depend on the loading rate, $\dot{\varepsilon}$, or the time to fracture, τ_f

$$E(0) = \frac{U_0}{\gamma\varepsilon_0} - \frac{\mathbf{k}\sum_m G_m\theta_m}{4G\gamma\varepsilon_0} \ln\frac{\dot{\varepsilon}_0}{\dot{\varepsilon}} \tag{53}$$

and

$$\sigma_f(0) = \frac{U_0}{\gamma} - \frac{k\sum_m G_m\theta_m}{4G\gamma} \ln \frac{\tau_f}{\tau_{0f}} \tag{54}$$

These results contradict the conclusions from the classic Arrhenius-type equations. Indeed, Eqs. (3), (5), and (9) yield: $\dot{\varepsilon}(T \to 0) = 0$, $E(T \to 0) = U_0/\gamma\varepsilon_0$, and $\sigma_f(T \to 0) = U_0/\gamma$.

Equations (52)–(54) show that finite rates of creep, relaxation of Young's modulus, and fracture at $T \to 0$ are conditioned by tunneling of the atoms through the potential barrier and the existence of zero-point vibrations. These phenomena are not taken into account in the classic Arrhenius-type equations. The tunneling contribution in the kinetics of the processes examined depends on the characteristic temperature.

To illustrate this conclusion, let us examine the E(T) dependence in Fig. 19, a. The peculiarities of the $F(\theta/T)$ function in the $T_t < T < T_b$ region result in the following expression:

$$E(T) \cong \frac{U_0}{\gamma\varepsilon_0} - k\frac{C_t\theta_t + G_b\theta_b}{2\gamma\varepsilon_0(G_t + G_b)} \ln\frac{\dot{\varepsilon}_0}{\dot{\varepsilon}} - \frac{2kG_tT}{C(G_t + G_b)} \ln\frac{\dot{\varepsilon}_0}{\dot{\varepsilon}} \tag{55}$$

The second term of Eq. (55) characterizes a decrease in modulus, caused by the atoms tunneling through the potential barrier because of the zero-point twisting and deformation vibrations; and the third term is that caused by thermally activated atoms overcoming the potential barrier by means of the energy of the twisting vibrations.

To calculate the tunnel effect, ΔE_{tnl}, on a modulus decrease when the temperature is raised in both temperature ranges, $T_t < T < T_b$ and $T_b < T < T_{melt}$, E(T) should be extrapolated to $T = 0$. The ordinate segment, ΔE_{tnl}, in Fig. 19, a is a measure of the tunneling. The decrease in modulus caused by thermal fluctuations appears in Fig. 19, a as ΔE_{fl}. ΔE_{tnl} seems to be of the same magnitude as ΔE_{fl}. This statement is valid over all temperature ranges from $T \cong 0$ K to $T \cong T_{melt}$. That is why the modified Eqs. (49)–(51) should be used to calculate the activation parameters from experimental data.

5.4 *Evaluation of the True Activation Parameters of Kinetic Equations*

The activation parameters of kinetics equations, as usually evaluated from experimental data, do not take into account the quantum statistics of atomic vibrations. In such case, one obtains the *apparent* activation parameters which have to be corrected in order to get the *true* ones.

To clarify this correction, let us consider the E(T) dependence in the $T_t < T < T_b$ range. The "classic" Arrhenius equation (5) results in:

$$\frac{\partial E}{\partial T} = -\frac{k}{\gamma_{ap}\varepsilon_0} \ln\frac{\dot{\varepsilon}_0}{\dot{\varepsilon}} \tag{56}$$

On the other hand, if a non-uniform energy distribution over vibrational modes is taken into account, then Eq. (5) yields

$$\frac{\partial E}{\partial T} = -\frac{k}{\gamma_{tr}\varepsilon_0}\frac{\partial F(\theta/T)}{\partial T}\ln\frac{\dot{\varepsilon}_0}{\dot{\varepsilon}} \tag{57}$$

In Eqs. (56) and (57), the "ap" and "tr" indices of the activation volume designate its "apparent" and "true" values, respectively.

Comparison between Eqs. (56) and (57) gives the correlation between the apparent and true activation volumes:

$$\gamma_{tr} = \gamma_{ap}\frac{\partial F(\theta/T)}{\partial T} \tag{58}$$

Similar considerations result in an expression for the connection between the apparent and true activation energies:

$$U_{0tr} = U_{0ap}\frac{\partial F(\theta/T)}{\partial T} \tag{59}$$

As can be seen from Eqs. (58) and (59), both parameters, γ_{ap} and U_{0ap}, increase with decreasing temperature; and at $T \to 0$, they become infinite: $\gamma_{ap} \to \infty$ and $U_{0ap} \to \infty$.

Now consider Dorn's method (see Sect. 5.1) of calculating both the apparent and true activation parameters.

Substitution of $F(\theta/T)$ for T in Eq. (28) results in

$$\gamma_{tr} = kF(\theta/T)\frac{\partial(\ln\dot{\varepsilon})}{\partial\sigma} \cong kF(\theta/T)\frac{\ln(\dot{\varepsilon}_2/\dot{\varepsilon}_1)}{\sigma_2 - \sigma_1}\bigg|_T \tag{60}$$

Comparison of the "quantum" Eq. (60) and "classical" Eq. (28) yields:

$$\gamma_{tr} = \gamma_{ap}\frac{F(\theta/T)}{T} \tag{61}$$

Similarly, the relation between the apparent and true activation energies is:

$$U_{0tr} = U_{0ap}\frac{F(\theta/T)}{T}. \tag{62}$$

Equations (61) and (62) show that both parameters decrease with decreasing temperature. As was shown in Sect. 5.1, if the activation parameters are evaluated by different methods, they demonstrate contradictory temperature dependences. Fig. 20 shows that, if the quantum statistics is taken into account, the true activation parameters are independent of temperature.

Whenever, the thermodynamic Debye temperature for polymers exceeds their condensed-state temperature. Therefore the quantum statistics of atomic vibrations should be taken into account. As the $F(\theta/T)$ function has no a precise analytical form, the true activation parameters cannot be evaluated precisely. Nevertheless, in the $T_b < T < T_{melt}$ range, only two vibrational modes, those with the ν_t and ν_b frequencies, are shown to give the greatest contributions in

$F(\theta/T)$ (see Sect. 5.2). If the contributions of the other modes are neglected, then – in the $T_b < T < T_{melt}$ region – $\gamma_{ap} \cong \gamma_{tr}$ and $U_{0ap} \cong U_{0tr}$.

To illustrate this statement, let us consider the $\sigma_f(T)$ dependences in Fig. 19, b. As concluded above, the change in the statistics of any vibrational mode increases its slope, $\partial\sigma_f/\partial T$. In the ultimate case, $\partial\sigma_f/\partial T$ tends to 90°. Therefore, a normal is dropped from the last experimental point (near $T \cong T_{melt}$) and the point at which it crosses the T axis is denoted by T_∞. If we can obtain the experimental data at $T > \theta_b/3$, then $\sigma_f(T)$ crosses the T axis at $T_b^* > T_\infty$. Hence, the *true* activation energy, U_0, drops in the interval between $U_{0\infty} = kT_\infty \ln(\tau_f/\tau_{0f})$ and $U_{0b} = kT_b^* \ln(\tau_f/\tau_{0f})$. For instance, for PE $U_{0b} \cong 115$ kJ mol^{-1} and $U_{0\infty} \cong 100$ kJ mol^{-1}, so that $U_0 = (100–115)$ kJ mol^{-1}. Consequently, $U_0 \cong U_{0b}$, and U_{0b} may be treated as a close approximation to U_0. A similar approach is used for calculation of the *true* activation volume. The calculated true activation parameters are listed in Table 1 and plotted in Fig. 20.

5.5 Physical Nature of the Activation Parameters of Kinetic Equations

Nowadays, the physical nature of the activation parameters (the activation energy, U_0, and the activation volume, γ) in the kinetic equations (3), (7), and (8) for the deformation rate, relaxation, and durability of polymers is being discussed intensively.

For oriented polymers, the U_0 parameter was observed to be close to the activation energy of the accumulation of free radicals and of new ends of the macromolecules. Both are caused by rupture of macromolecules [9, 10]. Therefore, U_0 was proposed to be directly related to the rupture of a chemical bond in the macromolecular skeleton. This proposal is confirmed by the numerical coincidence of U_0 with the activation energy of thermal destruction, U_{td}, as evaluated from mass spectrometry data [9]. Nevertheless, this interpretation of U_0 cannot explain the other experimental data:

(i) For polyolefines, the activation energy of fracture is equal to ca. 120 kJ mol^{-1}, whereas, the energy of bond dissociation is much higher for paraffins, (320–340) kJ mol^{-1} [107].
(ii) The equality $U_0 \cong U_{td}$ is fulfilled only in the initial stage of thermal destruction (when the fraction of the polymer destroyed is less than 1%), whereas in the further stages, $U_0 \cong 0.5U_{td}$ [108].
(iii) It was shown in Sect. 4.1 that the activation energy of fracture coincides with the activation energy of deformation. Moreover, this equality as well as the numerical values of energies holds for both oriented and isotropic polymers [13]. However, for isotropic polymers, the contribution of intermolecular interactions in the U_0 parameter is known to reach ca. 70% [14].

If U_0 is treated as the energy of rupture of a single chemical bond in a macromolecule, then the activation volume, γ, has to be of the same order of magnitude as the volume of a single chemical bond, γ_a. Yet, the experimental γ values are several powers often greater than γ_a. To explain this contradiction, the true stress, rupturing bond, Σ, was assumed to be q times larger than the applied stress σ: $q = \Sigma/\sigma$, from which $\gamma = q\gamma_a$ [9, 10]. This assumption clearly explains an increase in polymer strength during the orientational drawing. It was suggested that the latter distributes the stress more uniformly in the amorphous (disordered) parts of a polymer, and decreases the overstress coefficient, q. This conclusion agrees with the data indicating a change in structure of amorphous regions, obtained by IR spectroscopy and nuclear magnetic resonance [9].

Nevertheless, further careful investigations did not discern a rigid correlation between polymer strength and the stress distribution on amorphous regions [109]:

(i) An increase in segmental orientation was revealed to be accompanied by either increase or decrease in sample strength.
(ii) A higher concentration and orientation of *trans* isomers is often assigned to the sample with higher strength.
(iii) Concentration and deformation of the excited chemical bonds in some super-oriented, ultra-high-strength samples were observed in some super-oriented, ultra-high-strength samples were observed to be higher than those in samples with lower strength.

Besides, the interpretation of the γ parameter is in contradiction with its interpretation for isotropic polymers. In Ref. [13], the γ parameter was assumed to be 10^3 times larger than the volume of a single chemical bond. It is close to the volume of a Kuhn's segment, i.e. of a chain of several repeating units.

The present approach was developed by the authors to provide a possibility of describing both deformation and fracture as processes of linear physics.

The concept that kinetics of deformation, relaxation, and fracture are based on thermal fluctuations presents alternative interpretation of the physical nature of the activation parameters. In the proposed approach, the U_0 and γ parameters are characteristics of a complicated, multistage process. It consists of generation of the excited chemical bonds, their relaxation, generation of the elementary nuclei of deformation/fracture, their coalescence, and macroscopic deformation/fracture.

The first stage, generation of regions containing the excited bonds, was shown in Sect. 4.4 to occur at the lowest rate. Therefore, this stage controls the rate of deformation/fracture. It is essential that the excited bonds can be generated in the absence of external stress, and only be due to thermal action, in both oriented and isotropic polymers and even in a polymer melt (Sect. 4.3). Consequently, the U_0 and γ parameters should be treated as the activation energy and the activation volume of the generation of regions containing the excited bonds. As the vibrational amplitude in these regions is much larger than that in the equilibrium state (Sect. 4.2), these regions can each be considered as a kind of activated complex, as the term is used in physical chemistry [110].

The energy necessary to generate the regions containing the excited bonds was shown in Sect. 5.4 to be taken up mainly from the energies of vibrational modes for twisting and deformation. As is known from spectroscopy, the maximum frequencies, ν_t and ν_b, and the mode Grüneisen parameters, G_t and G_b, of these vibrations, depend strongly on intermolecular interactions. These vibrations are not localized in a given macromolecule but embrace large ensembles of jointly moving (correlated) macromolecular segments. Therefore, a correlated ensemble of macromolecules participates in the generation of the regions containing the excited bonds. Hence, both intra- and intermolecular interactions contribute to the values of the activation parameters. This is why the activation volume has a rather large value (of the order of $10^{-28}\,m^3$ as seen from Table 1).

Consequently, the activation parameters, U_0 and γ, should be attributed rather to the ensemble of correlatively moving macromolecules as the excited bonds are generated. The consequent processes (generation of nuclei of deformation/fracture, slippage of molecular chains, and many others) do not influence the value of the activation energy strongly, because their rates are much higher than the rate of the generation of the excited bonds.

5.6 *Relaxation Transitions in Polymers as Caused by a Change in the Statistics of Fundamental Vibrations*

Deviations from linearity in the temperature dependence of Young's modulus and of tensile strength are well known and are being investigated intensively by methods of relaxation spectroscopy [95, 111, 112]. Such peculiarities are usually treated as relaxation transitions. In this Section, a correlation between the characteristic temperatures, T_m, and the transition temperatures of relaxation, T_r, will be drawn.

In Sect. 5.2, two characteristic temperatures, $T_m = T_t$ and $T_m = T_b$, were shown to be similar to the temperatures of the main (β and α) relaxation transitions. At the same time, there are many more other minor relaxation transitions in a polymer spectrum, especially for isotropic polymers. As can be seen from Table 4, each relaxation transition temperature, T_r, corresponds to a certain characteristic temperature, T_m [113].

The similarity between T_m and T_r is not surprising. At $T \cong T_m$, the contribution of the *m*-th vibrational mode to the heat capacity, C_m, was shown (see Sect. 5.2) to abruptly increase from the minimum value ($C_m \cong 0$) to the maximum value ($C_m \cong k$). The heat capacity characterizes the ability of fundamental vibrations to absorb energy. Therefore, an increase in C_m at $T \cong T_m$ permits the generation of the thermal fluctuations that cause the relaxation transitions. Hence, the change in the statistics of any fundamental vibrational mode is assumed to determine the relaxation transition [34, 35, 40, 113].

Moreover, T_m is similar with T_0 in the Williams-Landel-Ferry equation or in the Fulcher-Vogel-Tamman equation [14]. The T_0 value is the lowest thermo-

Table 4. A comparison between the characteristic temperatures, T_m, and the relaxation transitions temperatures, T_r

Polymer	Maximum frequencies of fundamental vibrations, $\nu_{m\,max}$, (cm^{-1}) [83, 84]	Characteristic temperatures, T_m(K)	Relaxation transition temperatures T_r(K)
Nylon 6	184–204	90–100	110–140
	310–340	150–170	170–180
	650	310	310–330
PAN	280	140	140
	338	160	160
	565	270	330–340
	800	390	380–390
PE	190–210	90–100	120–130
	340	170	150–170
	500–600	240–290	220–270
	720–750	350–360	320–370
	288	110	110–140
	310–380	150–180	180–230
	535	255	250–300
POM	288	140	140–150
	470	230	260
	568	280	270–320
	844	410	400
PP	315	150	150–170
	470	230	230
	570	280	280–290
PTFE	345	170	160–170
	510	250	230–250
	560	270	270–280
	685	330	380–410

dynamic limit of the experimental T_m value. It can be achieved at infinitely slow change in temperature ($\dot{T} \to 0$) and/or applied stress ($\dot{\sigma} \to 0$). The change in the statistics of certain fundamental vibrational modes at $T \cong T_m$ probably, triggers the corresponding relaxation transition at $T \cong T_r \geq T_m$.

6 Conclusions

In this paper, the kinetics of deformation, relaxation, and fracture for perfectly oriented polymers over a wide temperature range are considered. The rates of these processes, as functions of temperature and applied stress, were shown to be described by Arrhenius-type equations. The latter have to be modified to take into account the non-uniform energy distribution over modes of atomic vibrations.

The kinetics of generation and evolution of the excited extended chemical bonds is also investigated. The activation energy for their generation was shown to be equal to that for macroscopic deformation and fracture. This result implies that the formation of elementary nuclei of deformation and fracture is caused by the relaxation of excited bonds. Therefore, the question: "Which process in a stressed solid is the first, local deformation or fracture?" has been disentangled at last. The initial process in a stressed solid is the generation of excited bonds, and its rate controls the rates of both deformation and fracture.

Creep rate, Young's modulus, and tensile strength are also shown to be connected with the mode parameters (the Grüneisen parameter and the maximum frequency) of the fundamental vibrations. We suggest, that this relation reflects the participation of different vibrational modes in the generation of excited bonds. Therefore, powerful energy fluctuations seem to play a decisive role in the deformation, relaxation, and fracture: in oriented polymers, their formation controls the kinetics of the macroscopic processes considered.

Acknowledgement. The authors wish to express their profound thanks to Prof. H.H. Kausch for his careful and useful remarks which have improved the manuscript.

7 References

1. Arrhenius S (1889) Z Phys Chem 4: 226
2. Becker R (1925) Z Phys 26: 919
3. Eyring H (1936) J Chem Phys 4: 283
4. Orowan E (1938) Proc Roy Soc London A 168: 307
5. Frenkel J (1955) Kinetic theory of liquids. Dover. New York
6. Krausz AS, Eyring H (1975) Deformation kinetics. Wiley, New York
7. Dorn JE (1957) Creep and recovery. ASTM, New York
8. Zhurkov SN (1965) Int J Fract Mech 1: 311
9. Regel VR, Slutsker AI, Tomashevskii EE (1974) Kinetic nature of the solids strength. Nauka, Moscow (in Russian)
10. Kausch HH (1987) Polymer fracture. 2nd Ed., Springer, Berlin
11. Petrov VA, Bashkarev AYa, Vettegren VI (1993) Physical background of durability prognosis for constructional materials. Politekhnika, St. Petersburg (in Russian)
12. Ward IM (1985) Adv Polym Sci 70: 1
13. Stepanov VA, Peschanskaya NI, Shpeizman VI (1984) Strength and relaxational phenomena in solids. Nauka, Leningrad (in Russian)
14. Bershtein VA, Egorov VM (1992) Differential scanning calorimetry of polymers. Ellis Harwood, Chichester
15. Tamuzh VP, Kuksenko VS (1978) Micromechanics of polymer materials fracture. Zinatne, Riga (in Russian)
16. Bronnikov SV, Vettegren VI, Korzhavin LN, Frenkel SYa (1983) Mekh Kompoz Mater Nr 5: 920
17. Bronnikov SV, Vettegren VI, Korzhavin LN, Frenkel SYa (1986) Vysokomol Soed A 28: 1963
18. Savelyev VD, Bronnikov SV, Vettegren VI (1988) Vysokomol Soed B 30: 83
19. Bronnikov SV, Vettegren VI, Korzhavin LN, Frenkel SYa (1988) Vysokomol Soed A 30: 2115
20. Vettegren VI, Bronnikov SV, Korzhavin LN, Frenkel SYa (1990) J Macromol Sci-Phys B 29: 285

21. Bronnikov SV (1991) Proc 8th Int Conf Deformation, Yield and Fracture of Polymers, Cambridge, p 48
22. Bronnikov SV, Vettegren VI, Frenkel SYa (1992) Proc 25th Europhys Conf Macromol Phys, St. Petersburg. Europhys Conf Abstr D 16: 240
23. Vettegren VI, Bronnikov SV, Frenkel SYa (1984) Vysokomol Soed A 26: 939
24. Bronnikov SV, Vettegren VI, Frenkel SYa (1993) Mekh Kompoz Mater 29: 446
25. Zhurkov SN, Vettegren VI, Korsukov VE, Novak II (1969) Proc 2nd Int Conf Fracture, Brighton, p 545
26. Vettegren VI, Novak II (1973) J Polym Sci, Polym Phys Ed. 11: 2135
27. Vettegren VI, Novak II, Friedland KJ (1975) Int J Fract Mech 11: 789
28. Vorobyov VM, Razumovskaya IV, Vettegren VI (1978) Polymer 19: 1267
29. Vettegren VI, Prokopchuk NR, Korzhavin LN, Frenkel SYa, Koton MM (1979) J Macromol Sci-Phys B 16: 163
30. Vettegren VI (1985) In: Steger WE (Ed) Progress in polymer spectroscopy. Proc 7th Europ Symp Polymer Spectroscopy, Teubner Verlag, Leipzig 9: 158
31. Vettegren VI, Friedland KJ (1975) Opt Spektr 38: 521
32. Vettegren VI (1984) Fiz Tverd Tela 26: 1699
33. Vettegren VI (1986) Fiz Tverd Tela 28: 3417
34. Bronnikov SV, Vettegren VI (1993) Proc 8th Int Conf Fracture, Kiev, p 356
35. Vettegren VI, Ibrogimov II, Bronnikov SV (1992) Proc 25th Europhys Conf Macromol Phys, St. Petersburg, Europhys Conf Abstr D 16: 230
36. Vettegren VI, Kusov AA (1982) Fiz Tverd Tela 24: 1598
37. Vettegren VI, Abdulmanov RR (1984) Fiz Tverd Tela 26: 3266
38. Gal AE, Vettegren VI, Perepelkin KE (1985) Vysokomol Soed B 27: 612
39. Vettegren VI (1987) Excited interatomic bonds and their contribution to fracture. Doctorate dissertation. Physical Technical Institute, St. Petersburg
40. Vettegren VI, Ibrogimov II, Bronnikov SV (1992) Proc 10th Europ Symp Polymer Spectroscopy, St. Petersburg, p A10
41. Bronnikov SV, Vettegren VI. Kalbina NS (1990) Mekh Kompoz Mater 26: 544
42. Bronnikov SV, Gromova ES (1993) Proc Fall '93 Meeting, Boston, p E8.2
43. Korsukov VE, Korelina NF, Petrov VA (1981) Fiz Tverd Tela 23: 387
44. Woods DW, Busfield WK, Ward IM (1984) Polymer (Comm) 25: 298
45. Gupta YM (1984) Polym Eng Sci 24: 851
46. Ishai O (1967) J Appl Polym Sci 11: 963
47. Waterman HA (1964) Kolloid-Z Z Polym 196: 1
48. Ricco T, Smith TL (1990) J Polym Sci, Polym Phys Ed. 28: 513
49. Lagakos N, Jarzynski J, Cole JH, Bucaro JA (1986) J Appl Phys 59: 4017
50. Haidar B, Smith TL (1990) Polymer 31: 1904
51. Mergenthaler DB, Pietralla M, Roy S, Kilian HG (1992) Macromolecules 25: 3500
52. Wolf F-P (1982) Colloid Polym Sci 260: 577
53. Joseph SH, Duckett RA (1978) Polymer 19: 837, 844
54. Assay JR, Lambertson DL, Guenter AH (1969) J Appl Phys 40: 1768
55. Zhurkov SN (1957) Vest AN SSSR 11: 78
56. Bronnikov SV, Vettegren VI, Kalbina NS, Korzhavin LN, Frenkel SYa (1989) Vysokomol Soed B 31: 627
57. Bronnikov SV, Vettegren VI, Kalbina NS, Korzhavin LN, Frenkel SYa (1990) Vysokomol Soed A 32: 1500
58. Roylance DK, De Vries KL (1971) J Polym Sci 9: 443
59. Wool RP (1980) Polym Eng Sci 20: 805
60. Wool RP (1986) J Polym Sci, Polym Phys Ed 24: 1039
61. Herzberg G (1945) Infrared and Raman spectra of polyatomic molecules Wiley, New York
62. Scott RA, Scheraga F (1966) J Chem Phys 34: 3054
63. Pechhold W (1971) J Macromol Sci, Rev Macromol Chem Phys 32: 123
64. Wool RP, Statton WO (1974) J Polym Sci, Polym Phys Ed 12: 1575
65. Novak II, Shablygin MV, Pakhomov PM, Korsukov VE (1975) Mekh Polim Nr 6: 1077
66. Pfeffer GA, Sumpter BG, Noid DM (1992) Polym Eng Sci 32: 1278
67. Bershtein VA, Egorov VM, Emelyanov YuA, Stepanov VA (1983) Polym Bull 9: 98
68. Stein RS (1964) SPE Trans 4: 178
69. Young RI (1973) J Mater Sci 8: 23

70. Hinton T, Rieder IG, Simpson LA (1974) J Mater Sci 9: 1331
71. Kuhn W (1934) Angew Chem 52: 289
72. Fuoss RM, Kirkwood JG (1941) J Amer Chem Soc 63: 385
73. Schatzki TF (1962) J Polym Sci 16: 337
74. Boyer RF (1987) In: Keinath SE, Miller RL, Reieke JK (Eds) Order in the amorphous "state" of polymers, Plenum Press, New York, p 135
75. Pechhold W (1968) Kolloid-Z Z Polym 228: 1
76. Pechhold W (1971) J Macromol Sci, Rev Macromol Chem Phys 32: 123
77. Gotlib YuYa, Darinskii AA (1969) Vysokomol Soed A 11: 2400
78. Gotlib YuYa, Darinskii AA (1970) Vysokomol Soed A 12: 2263
79. Gotlib YuYa, Darinskii AA, Svetlov YuE (1986) Physical kinetics of macromolecules, Khimiya, Leningrad (in Russian)
80. Robertson RE (1978) J Polym Sci, Polym Symp 63: 173
81. Sherby OD, Dorn JE (1958) J Mech Phys Solids 6: 145
82. Mitra SK, Osborne PW, Dorn JE (1961) Trans AIME 221: 1206
83. Trevino S, Boutin H (1967) J Macromol Sci-Chem A 1: 723
84. Safford GL, Naumann AW (1967) Adv Polym Sci 5: 1
85. von Dechant J, Danz R, Kimmer W, Scholke R (1972) Ultrarot-spektroskopische untersuchungen an polymeren. Akademie-Verlag, Berlin
86. Pauling L, Wilson EB (1935) Introduction to quantum mechanics. McGraw-Hill, New York
87. Gilman II (1968) J Appl Phys 39: 6068
88. Salganik RL (1972) Fiz Tverd Tela 12: 1336
89. Slutsker AI (1989) Makromol Chem, Macromol Symp 27: 207
90. Slutsker AI, Veliev TM, Alieva IK, Abasov SA (1991) Makromol Chem, Macromol Symp 41: 109
91. Leibfried G (1955) Gittertheorie der mechanischen und thermischen Eigenschaften der Kristalle. Springer, Berlin
92. Titenkov LS, Vettegren VI, Bronnikov SV, Zelenev YuV (1985) Vysokomol Soed B27: 857
93. Vettegren VI, Titenkov LS, Bronnikov (1992) J Therm Anal 38: 1031
94. Dole M (1960) Fortsch Hochpolymer-Forsch 2: 221
95. Wunderlich B, Baur H (1970) Adv Polym Sci 7: 151
96. Tarasov V (1950) Zh Fiz Khim 24: 111
97. Lifshitz IM (1952) Zh Eks Teor Fiz (Sov Phys JETP) 22: 475
98. Girifalco LA (1973) Statistical physics of materials. Wiley, New York
99. Tasumi M, Simanouchi T, Miyazawa T (1962) J Mol Spectr 9: 261
100. Zerbi G, Piseri L (1968) J Chem Phys 49: 3840
101. Bronnikov SV, Vettegren VI, Vorobyov VM, Korzhavin LN, Frenkel SYa (1984) Vysokomol Soed B 26: 380
102. Bronnikov SV, Vettegren VI, Korzhavin LN, Frenkel SYa (1984) Vysokomol Soed A 26: 2483
103. Bronnikov SV, Vettegren VI, Korzhavin LN, Frenkel SYa (1984) Vysokomol Soed A 31: 1264
104. Bronnikov SV, Vettegren VI, Korzhavin LN, Frenkel SYa (1984) Proc 13th Conf Mechanisms of Polymer Strength and Toughness, Prague, p 41
105. Bronnikov SV, Vettegren VI, Frenkel SYa (1992) Polym Eng Sci 32: 1204
106. Bronnikov SV, Vettegren VI, Frenkel SYa (1993) J Macromol Sci-Phys B 32: 33
107. Denisov ET (1971) Rate constants of homolytic reactions in liquids Nauka Moscow (in Russian)
108. Regel VR, Amelin AV, Pozdnyakov OF, Sanfirova TP (1973) J Polym Sci C 42: 1399
109. Vettegren VI (1992) Unpublished data
110. Eyring H, Lin SH, Lin SM (1980) Basic chemical kinetics. Wiley, New York
111. Boyer RF (1985) in: Petrick RA (Ed) Polymer yearbook, Harwood Acad Publ, New York, 2: 233
112. Bartenev GM, Barteneva AG (1992) Relaxation properties of polymers Khimiya, Moscow (in Russian)
113. Vettegren VI, Bronnikov SV, Ibrogimov II (1994) Vysokomol Soed 36: 1294

Revised manuscript received and
accepted by the editor May 31st, 1995

Ultrasonic Spectroscopy for Polymeric Materials

K. Matsushige[1], N. Hiramatsu[2] and H. Okabe[3]

[1]Faculty of Engineering, Kyoto University, Yoshida-honmachi, Sakyo-ku, Kyoto 606-01, Japan
[2]Faculty of Science, Fukuoka University, Nanakuma, Johnan-ku, Fukuoka 814-01, Japan
[3]Faculty of Engineering, Kyushu University, Hakozaki, Higashi-ku, Fukuoka 812-81, Japan

The principle of the novel ultrasonic spectroscopy utilizing a wide-band polymeric transducer and the FFT (fast Fourier transformation) analysis of a single pulse is introduced and its application to various polymeric materials is reviewed. Ultrasonic analysis of mechanical relaxation processes and phase transitions in solid polymers as well as practical non-destructive inspection of defects in composite materials are described.

Advances in Polymer Science Vol. 125

1 Introduction

Polymeric materials composed of flexible molecular chains respond very sensitively to changes in various surrounding conditions such as temperature, stress, pressure, and so on. For example, temperature variation induces in polymeric materials drastic changes in mechanical and electrical properties as well as the phase transitions of melting and crystallization. Since these phenomena are accompanied by considerable variation in viscoelastic characteristics and thus in density (ρ) and elastic modulus (E) of the materials, the ultrasonic method which can detect quite sensitively these viscoelastic properties becomes one of the most valuable and important methods for investigating the macro- as well as microscopic structural changes occurring in the materials.

Ultrasonic analysis of polymeric materials has been carried out from various viewpoints. One of present authors (KM) has performed precise measurements of ultrasonic velocity under tensile stress condition and utilized an acoustic emission (AE) phenomenon to investigate the formation of microscopic cracks and the fracturing process of polymers [1]. Instrumental advancements have also enabled us to conduct rapid and even two-dimensional ultrasonic analysis of polymeric materials including composite materials [2, 3].

In this article, we introduce a recently developed ultrasonic spectroscopy method and review its application to polymeric studies. First, the principle of this ultrasonic spectroscopy is explained including the instruments and data analysis methods. Then, actual application of this measuring system is described for the characterization of solid polymers and the observation of phase transition phenomenon in liquid crystals and ferroelectric (VDF/TrFE) copolymer. Finally, the extension of this system to two-dimensional measuring and the application to non-destructive testing of CFRP (carbon fiber reinforced plastics) are discussed.

2 Principle of Single Pulse Ultrasonic Spectroscopy

2.1 *Historical Background*

The frequency is considered an important parameter, providing much information about the detailed mechanism in phase transition phenomena and the characteristics of the flaws in nondestructive testings. Ultrasonic spectroscopy is analogous to optical spectroscopy except it uses ultrasonic waves instead of light waves. Therefore, the success of optical spectroscopy has actually stimulated the development of similar techniques in the ultrasonic field. A theoretical and experimental study of ultrasonic spectroscopy was started by Kräutkramer [4] and Gericke [5, 6]. They attempted to use this technique for nondestructive testing of the defects in metals [5–8]. However, in the ultrasonic field, it was very hard to generate a wide band signal because of the absence of a wide-band transducer. Consequently this technique was not in practical use for some time. Later the development of piezoelectric polymer films as wide-band transducers and the digital data analysis by computer enabled one to use this ultrasonic spectroscopic technique in practice, as described below.

2.2 Earlier Techniques for Ultrasonic Spectroscopy

Before the development of modern techniques, there was an interest in the frequency dependence of the ultrasonic response of materials. The earlier spectroscopic techniques were classified as follows.

2.2.1 Manual Frequency Variation

The oldest technique for obtaining the ultrasonic transmission characteristic of a material as a function of the frequency is manual tuning of a conventional pulse echo measuring instrument. Of course, tunable and wide-band transmitter and receiver are required to carry out this technique. However, a narrow-band transducer had to be selected according to the test frequency and consequently this technique is clumsy and rarely used outside the laboratory.

2.2.2 Frequency Modulation

Automatic tuning can be substituted for manual frequency variation. Instruments employing such frequency modulated signals have been in use for many years for measuring thickness by ultrasonic resonance. This resonance electric equipment usually does not provide a linear amplitude vs frequency readout and is therefore only suitable for the detection of resonance peaks in spectra.

2.2.3 Multiple Frequency

The technique uses two or more tuned conventional pulse echo electric sensors connected to a single wide-band transducer. The data obtained are displayed on a single cathode-ray tube by means of a multitrace procedure. However, the data has frequency as a parameter. This technique has the potent of "real-time" spectroscopy and is suitable for the observation of transient phenomena.

The disadvantage of this technique is larger size and higher cost of the equipment with increasing number of measuring frequencies. For this reason it has thus far been implemented only for two measuring frequencies.

2.2.4 Pulse Shape Examination

Another approach for ultrasonic spectroscopy is the examination of the shape of an ultrasonic pulse before and after transmission in a sample. A theoretical and experimental study of ultrasonic pulse distortion was conducted by Gericke [5], leading to the conclusion that certain changes in the pulse spectrum can be recognized by a visual examination of the pulse shape. However, relatively complicated changes in the spectrum of the ultrasonic pulse may not easily be detectable from the pulse shape, and it requires operator skill for interpreting the observed pulse shapes.

2.3 Single Pulse Excitation Method

The modern and realistic approach would be to record digitally and subsequently analyze numerically the ultrasonic pulse by means of Fast Fourier Transformation (FFT). The development of piezoelectric polymer films as wide-band transducers and digital data analysis by computer have enabled this technique to be used easily [9].

Figure 1 shows the basic diagram of this single pulse excitation method. In the figure, small letters represent the transfer functions in time domain and capital letters represent the frequency transfer function. As shown in Fig. 1, the observed function $x(t)$ may be described as the following convolution chain:

$$x(t) = \int g_2(t_3)\,dt_3 \cdot \int h(t_2)\,dt_2 \cdot \int g_1(t_1) \cdot p(t - t_1 - t_2 - t_3)\,dt_1 \,. \quad (1)$$

Here, $p(t)$ is the electric signal applied to the transducer at time t, $g_1(t)$ is the electro-acoustic conversion function of the transducer, $g_2(t)$ is the acoustic-electro conversion function of the transducer, and $h(t)$ are the response functions of the sample. The Fourier transformation of $p(t)$, $g_1(t)$, $g_2(t)$ and $h(t)$ are defined as follows:

$$P(\omega) = \int p(t) \cdot e^{-i\omega t}\,dt \quad (2)$$

$$G_1(\omega) = \int g_1(t) \cdot e^{-i\omega t}\,dt \quad (3)$$

$$G_2(\omega) = \int g_2(t) \cdot e^{-i\omega t}\,dt \quad (4)$$

$$H(\omega) = \int H(t) \cdot e^{-i\omega t}\,dt \quad (5)$$

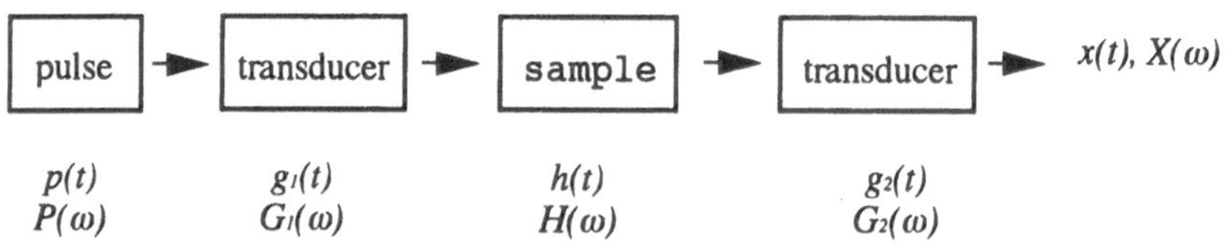

Fig. 1. Concept of the ultrasonic spectroscopy

$$X(\omega) = \int x(t) \cdot e^{-i\omega t} dt \tag{6}$$

and the invert FFTs of previous functions are described as follows:

$$p(t) = \frac{1}{2\pi} \int P(\omega) \cdot e^{-i\omega t} d\omega \tag{7}$$

$$g_1(t) = \frac{1}{2\pi} \int G_1(\omega) \cdot e^{-i\omega t} d\omega \tag{8}$$

$$g_2(t) = \frac{1}{2\pi} \int G_2(\omega) \cdot e^{-i\omega t} d\omega \tag{9}$$

$$h(t) = \frac{1}{2\pi} \int H(\omega) \cdot e^{-i\omega t} d\omega \tag{10}$$

$$x(t) = \frac{1}{2\pi} \int X(\omega) \cdot e^{-i\omega t} d\omega \,. \tag{11}$$

Since $x(t)$ is the convolution of $p(t)$, $g_1(t)$, $g_2(t)$ and $h(t)$ as shown in Eq. (1), the frequency transfer function $X(\omega)$ is described with $P(\omega)$, $G_1(\omega)$, $G_2(\omega)$ and $H(\omega)$ by:

$$X(\omega) = P(\omega) \cdot G_1(\omega) \cdot H(\omega) \cdot G_2(\omega) \,. \tag{12}$$

Therefore, if we have the frequency transfer function of input pulse and transducer, we can calculate the frequency transfer function of sample $H(\omega)$ by using the Fourier transformation of measuring function $x(t)$. Furthermore, if it is possible to get the next transfer function, which is regarded as an instrumental function, for the same system but without sample,

$$X'(\omega) = P(\omega) \cdot G_1(\omega) \cdot G_2(\omega) \tag{13}$$

we can calculate $H(\omega)$ from:

$$\frac{X(\omega)}{X'(\omega)} = H(\omega) \,. \tag{14}$$

In many cases, the measuring data for a "standard" sample provide the frequency transfer function $X'(\omega)$, and one can get the desired function of $H(\omega)$.

To carry out this procedure, a wide frequency characteristic of the signal is required. In this pulse excitation method, a rectangular pulse like the δ-function is used. When height and duration of the pulse are H and T respectively, the spectrum of the pulse (Fig. 2) is

$$P(\omega) = \frac{HT}{2\pi} \frac{\sin\left(\frac{\omega T}{2}\right)}{\left(\frac{\omega T}{2}\right)} \,. \tag{15}$$

The intensity of the spectrum has the maximum value of $HT/2\pi$ at a frequency of

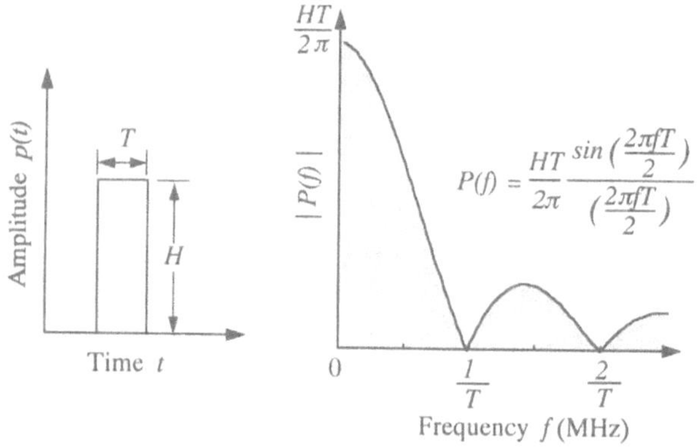

Fig. 2. Rectangular pulse and its spectrum

0 Hz and decreases quickly with a frequency period of 1/T Hz. Therefore, the measurements should be performed in a frequency range lower than 1/T Hz and it is desirable to generate the pulse with a short duration T and higher amplitude H.

As described above, the single pulse excitation method is very simple and powerful, but to put this technique into practice we have had to wait for the development of the wide-band transducer and the digital electric instruments described below.

2.4 Instruments for Single-Pulse Ultrasonic Spectroscopy

2.4.1 Configuration

The basic configurations are divided into two types according to whether the ultrasonic waves undergo transmission or reflection (Fig. 3). The sharp electric pulse with broad frequency components produced by a pulse generator is fed to a transducer. The ultrasonic waves transmitted through the sample or reflected by the surface of the sample are reconverted to electric signals by the same or facing transducer. After amplification, the signals are digitized and recorded using high speed wave memory, and then transferred to a computer. After the measurements, the data stored in the computer memory are analyzed using the FFT (fast Fourier transformation) technique. The power spectrum and relative phase shift thus calculated provide information on the frequency dependence of ultrasonic attenuation and velocity of the tested sample, respectively. Therefore, just one ultrasonic pulse is enough to obtain data for a whole frequency spectrum.

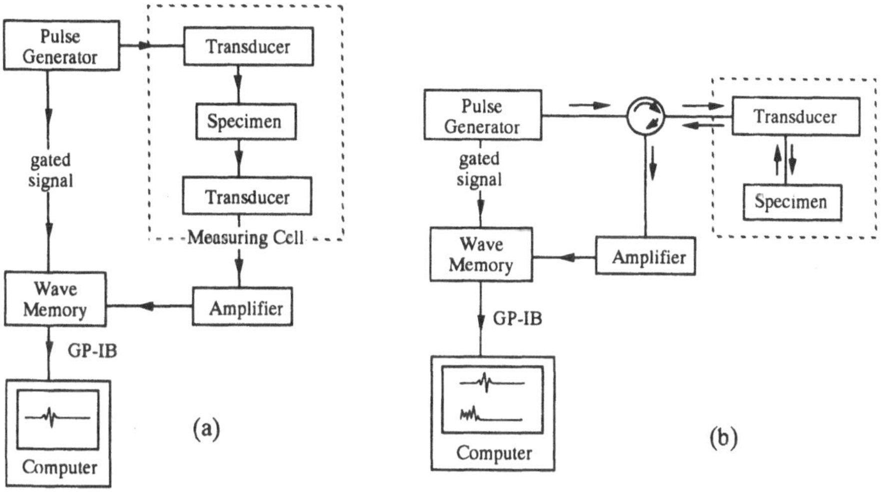

Fig. 3a, b. Basic configurations of measuring system: (**a**) transmission type; (**b**) reflection type

2.4.2 Transducers

The most important and difficult thing in the construction of the ultrasonic spectroscopy measuring system is the selection of the transducers, because they are required to generate and receive the wide-band ultrasonic waves with a flat frequency characteristic and high efficiency. The most commonly used transducer has been either a highly damped ceramic or a lithium sulfate transducer but its available frequency range is too narrow for this spectroscopy. However, recently developed piezoelectric polyvinylidene fluoride (PVDF) or copolymer of vinylidene fluoride-trifluoroethylene (VDF/TrFE copolymer) films have extremely wide and flat-frequency characteristics and low acoustic impedance comparable to that of water [10]. The characteristics of PVDF films are shown in Table 1, compared with those of other transducers. For example, the PVDF transducer that is 30 mm thick can be used to emit ultrasonic waves below 25 MHz. Furthermore, it is easy to produce from it a transducer that has a large area, and its flexibility is useful in producing a spherical surface for a focus type transducer. For the characteristics described above, the application of these polymer transducers to various fields such as medical and hydro-transducers is expanding rapidly. However, it should be remembered that the piezoelectric activity of these polymeric transducers is not thermally stable and is easily degraded when they are heated to around 80 °C, and so special care must be taken in the construction of measuring cells, as explained below.

2.4.3 Electric Equipment

Modern electric equipment has sufficiently wide frequency characteristics compared with the transducers. For example, with the fastest wave memory (50 GHz

Table 1. The characteristics of materials used for the transducers and sound media

	Density (10^3 kg/m^3)	$\varepsilon/\varepsilon_0$	Piezoelectricity (pC/N)	Sound Velocity (km/s)	Acoustic Impedance (10^6 Pa·s/m)	Mechanical Quality Factor	Electromechanical Coupling Factor (%)
PVDF	1.79	12	20	1.4	2.5	5	20
Quartz	2.65	4.5	2	5.7	15	$< 10^5$	11
Rochelle Salt	1.77	350	275	3.2	5.6	100	73
PZT	7.5	1400	300	3.2	24	1000	68
Water (0 °C)	1.0	80	—	1.4	1.4	—	—
Air (0 °C)	1.29×10^{-3}	1	—	0.33	4.27×10^{-4}	—	—

sampling), it is possible to evaluate the characteristic of a sample below about 10 GHz. And the speed and capacity of modern microcomputer are enough for the Fourier data analysis of the ultrasonic spectroscopy.

2.5 Data Analysis

2.5.1 Attenuation

Since ultrasonic attenuation of samples is defined as the inverse of $H(\omega)$ defined by Eq. (5), frequency dependence of attenuation can be examined by comparing two power spectra obtained with and without the samples (Eq. (14)). However, there is a problem with reflection at the surface of a sample, and so, in the case of a solid sample, frequency dependence of attenuation is examined by comparing two power spectra obtained from samples of different thicknesses to overcome this problem. The signal amplitude A at frequency f after transmitting through the sample is expressed by Eq. (16) for a sample of t_1 thickness and by Eq. (17) for one of t_2 thickness [11]:

$$A_1(f) = (1 - R)^2 A_0 \exp(-a(f)\cdot t_1) \tag{16}$$

$$A_2(f) = (1 - R)^2 A_0 \exp(-a(f)\cdot t_2) \tag{17}$$

where R is the reflectivity, A_0 is the amplitude before entering the sample, and a(f) is the attenuation coefficient. From Eqs. (16) and (17), a(f) is finally expressed as

$$a(f) = \frac{1}{(t_2 - t_1)} \ln\left(\frac{A_1(f)}{A_2(f)}\right). \tag{18}$$

Thus, the attenuation coefficient can be plotted as a function of frequency by calculating the logarithm of the ratio of the power spectra obtained from two samples of different thicknesses.

2.5.2 Sound Velocity

Wave velocity can be calculated from the phase of $H(\omega)$ defined by Eq. (5) and the information of the propagating length, and the phase $\theta(\omega)$ of $H(\omega)$ can be calculated from

$$\theta(\omega) = \tan^{-1}\left(\frac{I(\omega)}{R(\omega)}\right) \tag{19}$$

where $R(\omega)$ and $I(\omega)$ are a real part and an imaginary part of the $H(\omega)$, respectively. The shift of phase by a sample is described by Eq. (20) with θ_0, that is the phase measured without a sample:

$$d\theta(\omega) = \theta(\omega) - \theta_0(\omega)\,. \tag{20}$$

Therefore, the sound velocity c(ω) of a sample is calculated from

$$c(\omega) = \frac{d_s}{\frac{d\theta(\omega)}{2\pi} \cdot \frac{1}{f}} = \frac{d_s}{\frac{d\theta(\omega)}{\omega}} . \tag{21}$$

Since the range of phase θ that can be calculated numerically is in the region of $-\pi \sim +\pi$, the phase shift dθ is corrected by addition of 2π, as the data lay almost on a smooth curve in a frequency domain as shown in Fig. 4.

2.5.3 Nondestructive Testing and Thickness Measurements

Ultrasonic correlation analysis in frequency (Fourier transformation of frequency) domain analysis was utilized to measure a thickness of the sample and to image the structure of the material. This technique comprises four processes: (1) calculation of the spectrum, (2) division by the power spectrum of a pulse or other component, (3) Fourier transformation into the "frequency" domain, and (4) analysis and imaging in the frequency domain. Here we obtain much higher resolution in the imaging and thickness measurements by applying the echo analysis developed in earthquake theory [12] and the thickness measurements methods for a thin layer [13, 14].

If there exists some acoustic mismatching at surfaces and/or interfaces, the ultrasonic waves are reflected to some degree at these interfaces and they may be described by the following convolution chain [13]:

$$a(t) = \sum_i \int g(t_5) \int h_1(\mathbf{r}_i, t_4) \int k(\mathbf{r}_i, t_3) \int h_2(\mathbf{r}_i, t_2) \times \int g(t_1) e(t - t_1 - t_2 - t_3 - t_4 - t_5) dt_1 dt_2 dt_3 dt_4 dt_5 . \tag{22}$$

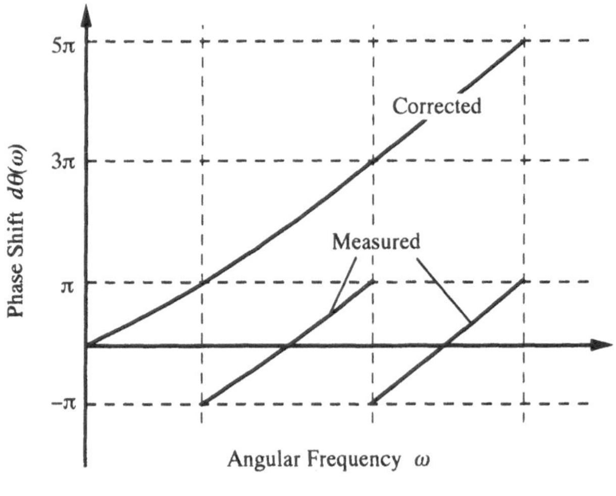

Fig. 4. Phase shift correction in frequency domain

Here, e(t) is the electric signal applied to the transducer at time t, g(t) is the electro-acoustic conversion function of the transducer, $\mathbf{r}_i$ is a position vector of the surfaces or interfaces, $h_1(\mathbf{r}_i, t)$ and $h_2(\mathbf{r}_i, t)$ are the response functions (diffraction and acoustic attenuation) of the medium between the transducer and the surface, and $k(\mathbf{r}_i, t)$ is the reflecting function. The sum is taken over all the surfaces and interfaces.

For simplification, we assume here only two interfaces as shown in Fig. 5. Then

$$\begin{aligned} a(t) &= \int g \int h_1(r_1) \int k(\mathbf{r}_1) \int h_2(\mathbf{r}_1) \int g \cdot e \, dt \\ &\quad + \int g \int h_1(\mathbf{r}_2) \int k(\mathbf{r}_2) \int h_2(\mathbf{r}_2) \int g \cdot e \, dt \\ &= \int g \int h_1(\mathbf{r}_1) \int k(\mathbf{r}_1) \int h_2(\mathbf{r}_1) \int g \cdot e \, dt \\ &\quad + \int g \int h_1(\mathbf{r}_1) \int \Delta h_1(dr) \int k(\mathbf{r}_2) \int \Delta h_2(dr) \int h_2(\mathbf{r}_1) \int g \cdot e \, dt \,, \end{aligned} \tag{23}$$

where informal expressions of the integral functions and parameters are used.

If surfaces 1 and 2 are located very close to each other and the space between the two surfaces is regarded as a homogeneous medium (for example, the two surfaces are parts of one layer (low-density region) as shown in Fig. 5, $\Delta h_1(\mathbf{dr}, t)$ and $\Delta h_2(\mathbf{dr}, t)$ are expected to have no frequency dispersion and to become the

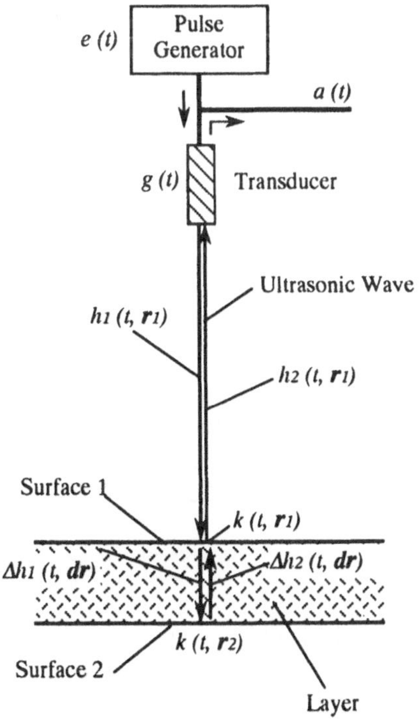

Fig. 5. Schematic view of ultrasonic path and its transfer function

same constant (α). Furthermore, if surfaces 1 and 2 are of the same type as shown in Figs. 2–5, the reflecting function $k(\mathbf{r}_i, t)$ is expected to have the same frequency dispersion

$$k(\mathbf{r}_2, t) = \beta k(\mathbf{r}_1, t) \quad (\beta; \text{ constant}). \tag{24}$$

Thus, Eq. (23) is derived as follows:

$$\begin{aligned} a(t) = &\int g \int h_1(\mathbf{r}_1) \int k(\mathbf{r}_1) \int h_2(\mathbf{r}_1) \int g \cdot e\, dt \\ &+ \int g \int \alpha h_1(\mathbf{r}_1) \int \beta k(\mathbf{r}_1) \int \alpha h_2(\mathbf{r}_1) \int g \cdot e(t - t_1 \ldots - 2\Delta t)\, dt, \end{aligned} \tag{25}$$

where Δt is transmitting time through $\mathbf{dr}$. We replace the integral in Eq. (25) with $x(t)$ as follows:

$$x(t) \equiv \int g \int h_1(\mathbf{r}_1) \int k(\mathbf{r}_1) \int h_2(\mathbf{r}_1) \int g \cdot e\, dt$$

$$a(t) = x(t) + \gamma\, x(t - t). \tag{26}$$

This type of signal is commonly treated in seismology [11], where $\gamma(\equiv \alpha^2\beta)$ and $\tau(\equiv 2\Delta t)$ represent echo ratio and ragtime, respectively. If $A(f)$, $G(f)$, $H_1(f)$, $K(f)$, $H_2(f)$, $E(f)$ and $X(f)$ are transfer functions at the frequency f, related to the respective response functions (Fourier transformations) of $a(t)$, $g(t)$, $h_1(t)$, $k(t)$, $h_2(t)$, $e(t)$ and $x(t)$, then

$$X(f) = G(f)^2 \cdot H_1(f) \cdot K(f) \cdot H_2(f) \cdot E(f) \tag{27}$$

$$\begin{aligned} |A(f)|^2 &= |\int \{x(t) + \gamma x(t - \tau)\} e^{-i2\pi ft} dt|^2 \\ &= |\int x(t) e^{-i2\pi ft} dt + \gamma e^{-i2\pi f\tau} \int x(t - \tau) e^{-i2\pi f \Phi^{tO\tau} \Psi} dt|^2 \\ &= |X(f) + \gamma e^{-i2\pi f\tau} X(f)|^2 \\ &= |X(f)|^2 |1 + \gamma(\cos 2\pi f\tau + i \cdot \sin 2\pi f\tau)|^2 \\ &= |X(f)|^2 (1 + \gamma^2 + 2\gamma \cos 2\pi f\tau) \end{aligned}$$

$$\frac{|A(f)|^2}{|X(f)|^2} = (1 + \gamma^2 + 2\gamma \cos 2\pi f\tau). \tag{28}$$

Because $|A(f)|^2/|X(f)|^2$ has the term $\cos 2\pi ft$, Fourier transformation of $|A(f)|^2/|X(f)|^2$ (in the frequency domain) has a peak at τ. Therefore, if the sound velocity is known, we can calculate the location (depth) or the thickness of the layer from the values τ. Furthermore, a two-dimensional scanning technique is successfully utilized to image the distribution of the layer in the medium as described in Sect. 4.

3 One-Dimensional Analysis of Solid Polymers

In this section, several results obtained by one-dimensional analysis using the single-pulse ultrasonic spectroscopic system are described, especially for the measurements of physical properties of organic compounds, such as polymeric materials and liquid crystals.

3.1 Characterization of Solid Polymers

The characterization of polymers, such as identification of the kind of polymeric material, is essentially important, but not easily conducted in general. If the polymers respond quite differently to ultrasonic waves, the spectroscopic information may be utilized as conventional data to identify the kinds of polymers tested. Here, as one of the simplest cases, the application of ultrasonic spectroscopy to the identification of the kind of polymer is demonstrated.

Figure 6 shows the measuring cell used for single-pulse ultrasonic spectroscopy carried out on various polymeric materials. Tested sheet-like specimens are set between transmitting and receiving transducers, and the surrounding space is filled with water as a coupling medium. The cell can be heated to 55 °C. Here, data are shown for four kinds of polymeric material – polyethylene (PE), polyacetal (PA), acrylonitrile butadiene styrene copolymer (ABS) and poly(methyl methacrylate) (PMMA). To extract the ultrasonic characteristics only from these samples, frequency dependence of attenuation was observed for samples of different thicknesses, data analysis being based on Eqs (16)–(18).

Figure 7 shows the time domain data for PE of 5 mm thickness, which is typical of the waveforms obtained from the polymers examined. The data were sampled at the rate of 2ns/word, which is fast enough to memorize the waveform in the MHz range. Similar waveform data were obtained for the PE sample with a thickness of 2 mm. Then the power spectra for these time domain data were calculated using FFT analysis, and those for PE of 2 mm and 5 mm thicknesses are shown in Fig. 8, revealing broad frequency components up to 12MHz and the maximum amplitude at different positions.

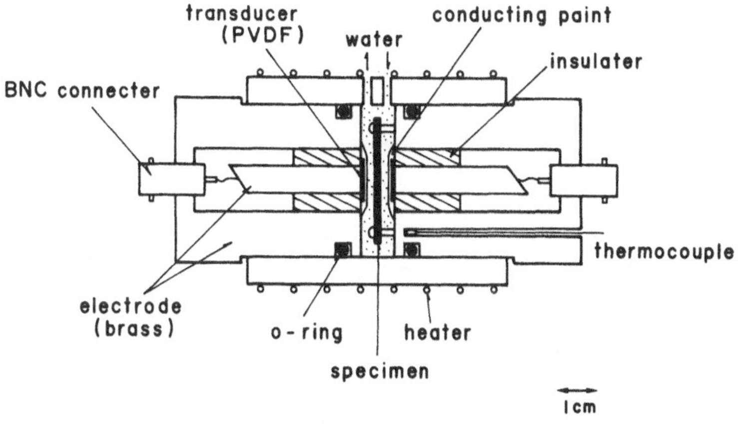

Fig. 6. Structure of a measuring cell

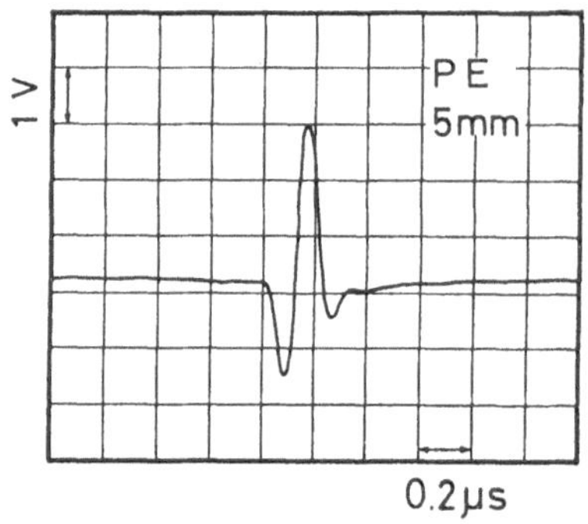

Fig. 7. Time domain signal of polyethylene (5 mm in thickness)

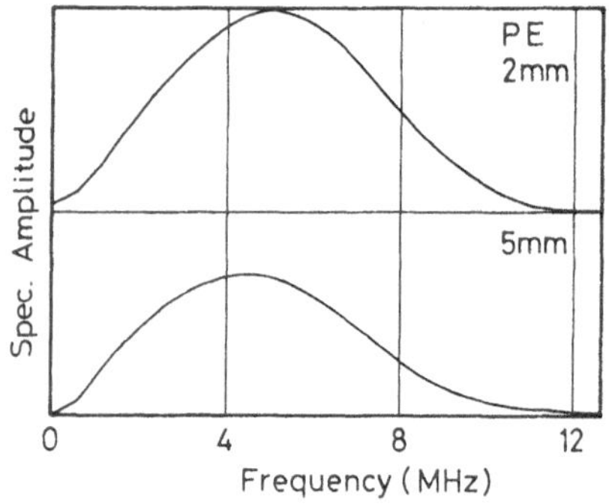

Fig. 8. Power spectra of the data obtained for polyethylene samples with thicknesses of 2 mm and 5 mm

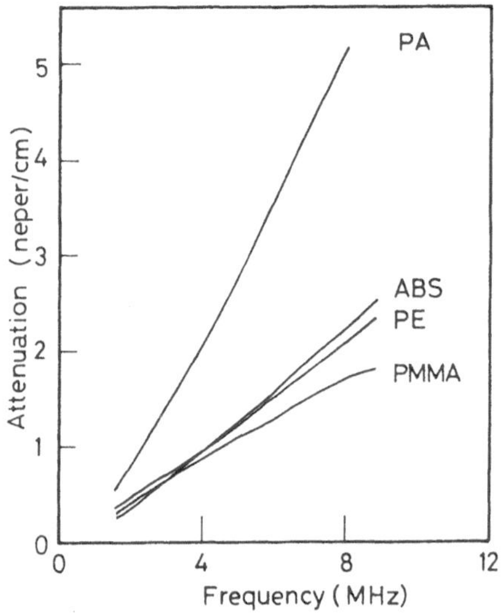

Fig. 9. Frequency dependencies of attenuations of four kinds of polymers

Next, the attenuation coefficient, a(f) was calculated using Eq. (18), being plotted as a function of frequency in Fig. 9. In the figure, the attenuation coefficients, not only for PE but also those for the other three polymers, are plotted as a function of frequency. It is noted that the attenuations increase almost linearly as the frequency becomes higher, suggesting that the attenuation in these polymers and in the frequency range examined is attributable to a hysteresis mechanism rather than a relaxation one. As demonstrated here, the polymeric materials with different chemical structures reveal quite different frequency dependency in the ultrasonic attenuation coefficient, and this fact suggests that this single-pulse ultrasonic spectroscopy measuring system can be utilized to identify the kinds of unclassified polymeric materials by comparing their ultrasonic characteristics to stored data from known polymers.

3.2 Phase Transition Phenomenon in Liquid Crystals

With phase transition, materials reveal quite drastic changes in various physical properties including elastic characteristics. Therefore, this one-pulse ultrasonic spectroscopy is expected to provide considerable information on the phase transition phenomena. The ultrasonic measuring system was used to investigate the nematic-isotropic phase transition in a typical liquid crystal, MBBA (p-methoxybenzylidene-p-n-butylaniline) in the temperature range from 30 to 55 °C [9]. In this experiment, the measuring cell shown in Fig. 10 was used. In the cell, the sample in a liquid state filled the all space between the transmitting and receiving PVDF transducers.

The waveforms transmitted through the sample changed considerably with increasing temperature, as shown in Fig. 11, reflecting the state of the sample. Figure 12 shows the power spectra of the transmitted waves observed at various temperatures, where the amplitude is expressed on a linear scale. As the temperature approaches the phase transition of this MBBA sample (about 47 °C), the intensity of the transmitted wave was observed to decrease gradually, and then to recover very rapidly above the phase transition temperature. This fact suggests that the absorption of the ultrasonic waves increases significantly when the sample undergoes phase transition accompanying density fluctuation.

This suggestion is clearly supported when the temperature changes in the ultrasonic absorption spectra are calculated from the transmission spectra, as shown in Fig. 13. When approaching the phase transition point, the absorption spectra grows considerably. In addition, it is noteworthy that the maximum frequency in the absorption spectra causes a shifting to a lower frequency. This fact suggests that the correlation length related to the density fluctuation becomes longer when the sample state becomes the phase transition point. Thus, it should be emphasized that this kind of information can be obtained only from the spectroscopic data, not by conventional ultrasonic measurements using a single frequency.

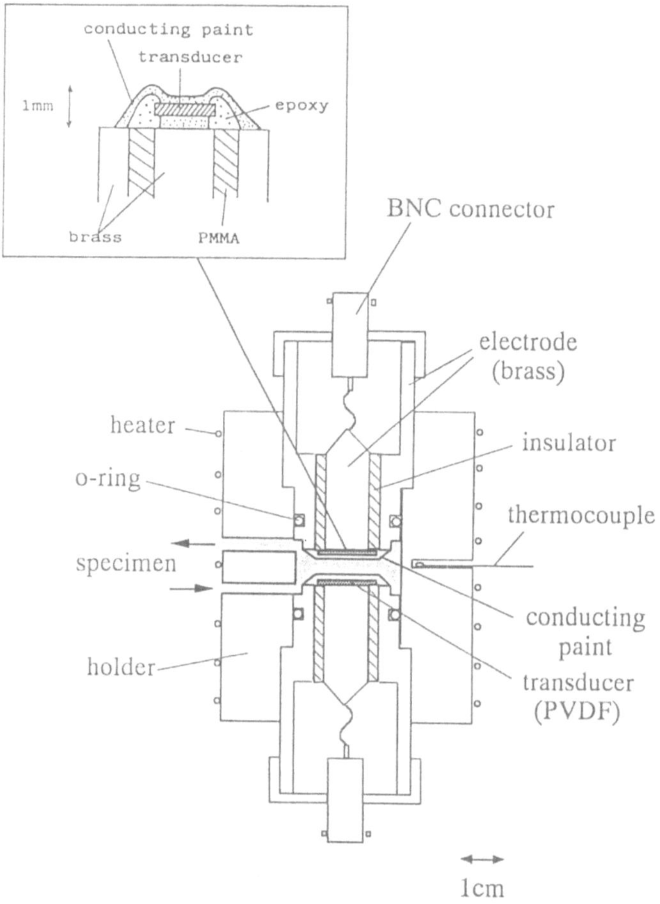

Fig. 10. Measuring cell for liquid crystalline samples

3.3 Ferroelectric Phase Transition in (VDF/TrFE) Copolymer

A copolymer of vinylidene fluoride-trifluoroethylene (VDF/TrFE) copolymer is well known as the polymer for which a clear Curie point was found for the first time in an organic material. At this Curie point, the polymer undergoes a solid-to-solid phase transition from paraelectric to ferroelectric phases with decreasing temperature. Therefore, the changes in the physical properties such as crystal structure, electrical and thermal properties upon the ferroelectric phase transition have drawn many researchers' interest. Here, the results concerning the ultrasonic spectroscopic investigation on acoustic and viscoelastic behaviour around the ferroelectric phase transition region of this copolymer are described [15].

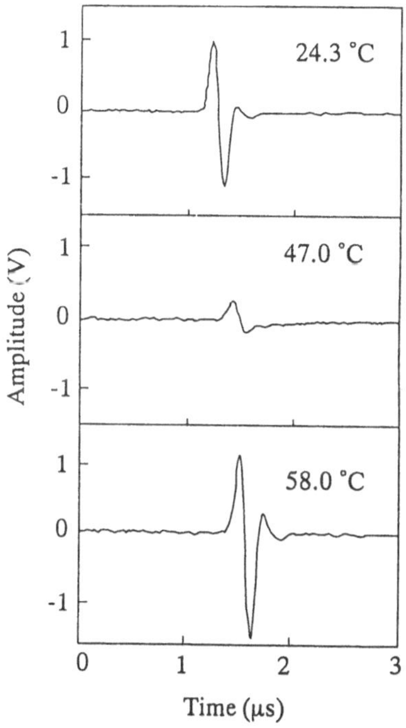

Fig. 11. Ultrasonic waveforms of MBBA at three different temperatures

The measurements were performed with the measuring cell shown in Fig. 14. This cell is designed for high temperature observation up to 100 °C. A polyimide was used as a buffer which prevents thermal deterioration of the transducers. Figure 15 shows the ultrasonic waveforms recorded by the transient recorder at three different temperatures, 30, 66 and 90 °C, which correspond to ferroelectric phase, transition region, and paraelectric phase, respectively. The lines shown in the upper part of the figure are the signals obtained without the sample, while those in the lower part are with the sample, where triggering points are located at 18 μs before t = 0 μs. The signals without the sample are almost identical, revealing the very stable character of the polyimide used as the buffer material. The sound velocity of polyimide is calculated to be about 2550 m/s and its variation due to a temperature change of about 60 °C is less than 1%. The position of wave signals transmitted through the sample shifts toward the right positions (later) and the shape of the waves shows a broadening with increasing temperature, while the amplitude reaches a minimum at 66 °C. Furthermore, it is obvious that interference waves such as those reflected at the specimen surfaces are very weak and can be neglected because of acoustic impedance matching between the polyimide buffers, transducers and sample.

Figure 16 shows the temperature and frequency dependence of ultrasonic absorption, which were obtained by FFT analysis of the ultrasonic signals observed at various temperatures. It is obvious that the absorption reaches a maximum around the ferroelectric phase transition of about 66 °C. Based on

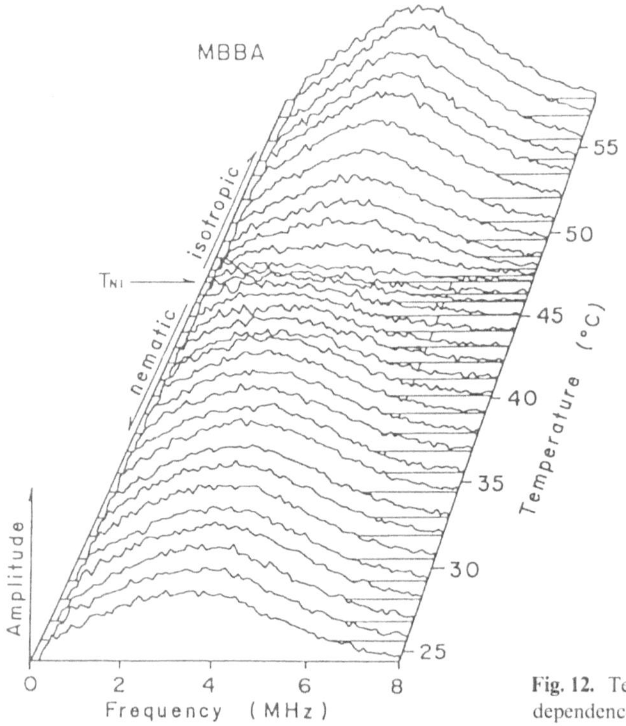

Fig. 12. Temperature and frequency dependence of absorption in MBBA

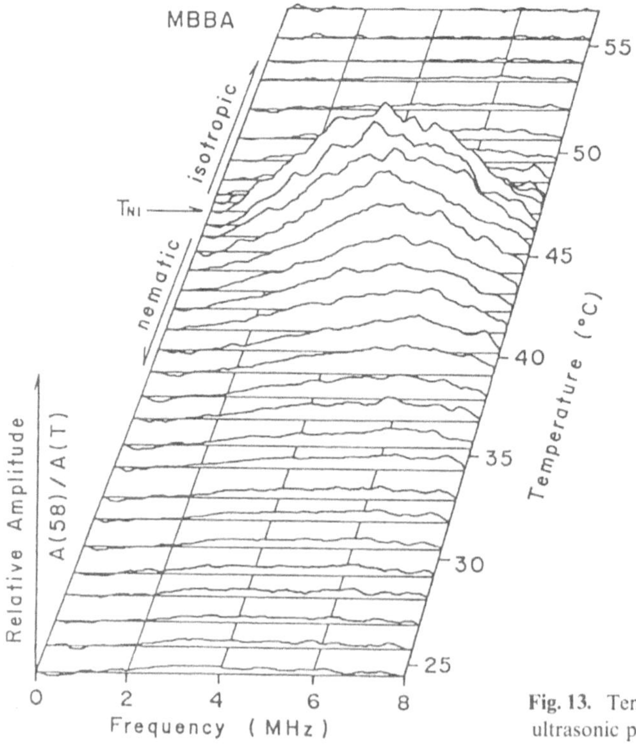

Fig. 13. Temperature variation in the ultrasonic power spectra of MBBA

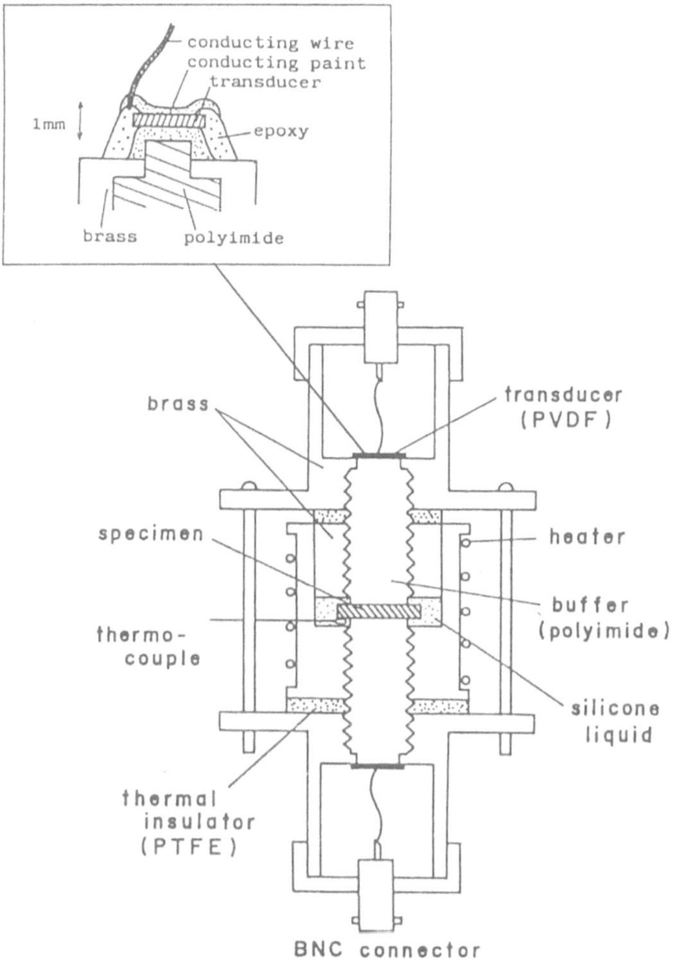

Fig. 14. Measuring cell for the observation of solid-to-solid phase transition

these data, one can investigate the details of the acoustic variations upon the phase transition. Maximum absorption temperatures vary slightly with frequency, as revealed in Fig. 17. Figure 18 plots the frequency dependence of the peak temperature, where the absorption reaches a maximum. The fact that the peak temperature decreases with increasing frequency suggests that the ferroelectric phase transition of the tested copolymer has, to some degree, the characteristic of second order phase transition.

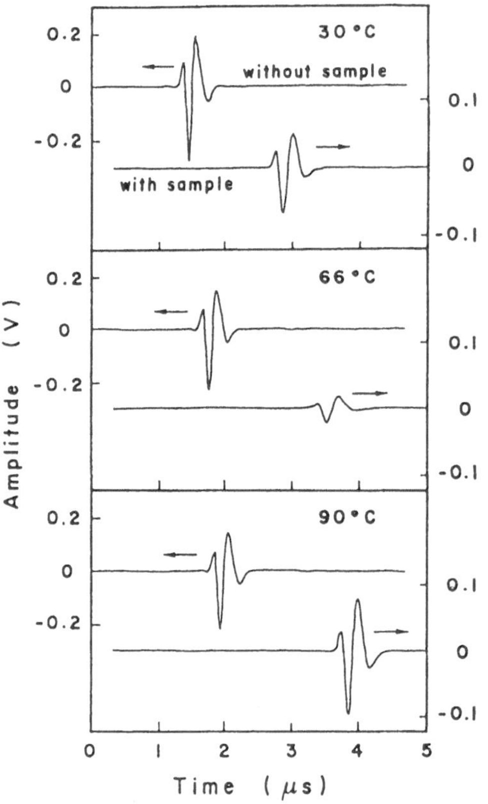

Fig. 15. Ultrasonic waveforms observed for P(VDF/TrFE) copolymer at three different temperatures below, around, and above the ferroelectric phase transition point

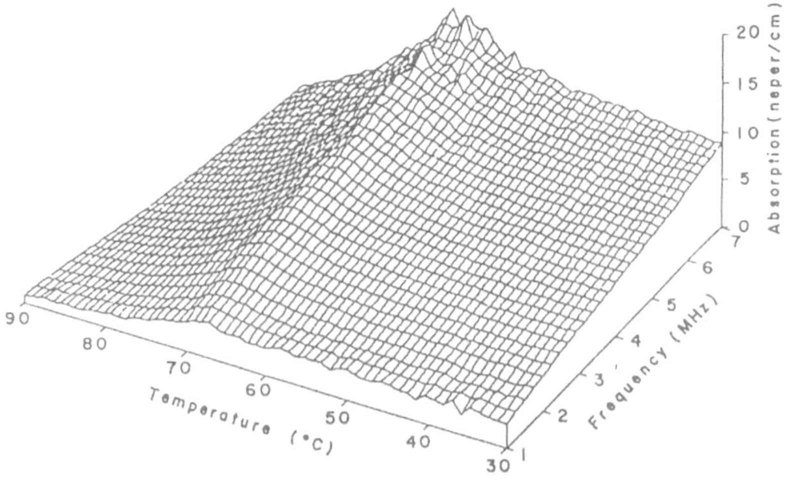

Fig. 16. Temperature and frequency dependence of absorption in P(VDF/TrFE) copolymer at the ferroelectric phase transition region

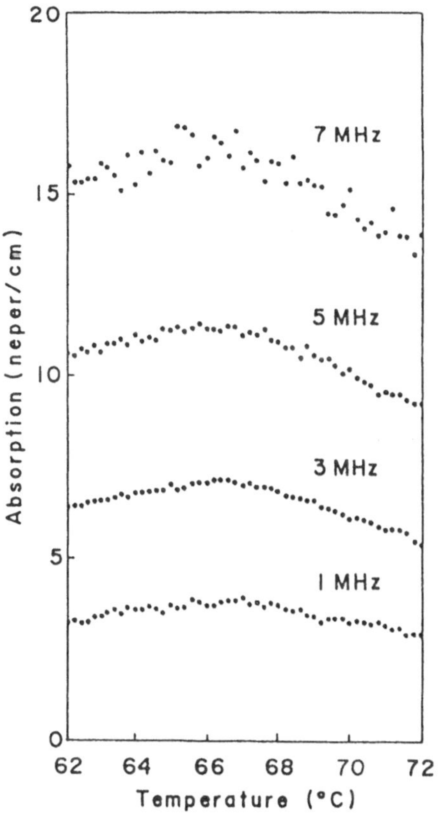

Fig. 17. Temperature dependence of absorption at several frequencies at the ferroelectric phase transition region of P(VDF/TrFE) copolymer

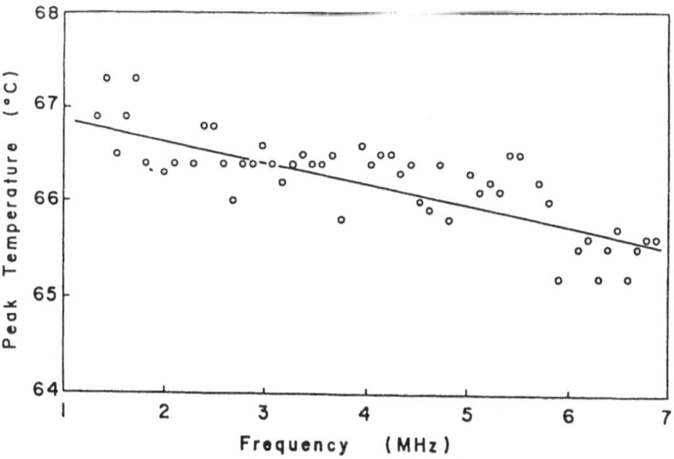

Fig. 18. Frequency dependence of the absorption peak temperatures of P(VDF/TrFE) copolymer

Figure 19 shows the frequency dependence of ultrasonic absorption at three typical temperatures. The data at the ferroelectric (30 °C) and transition regions (66 °C) are located almost on straight lines, suggesting that the absorption in and near the ferroelectric phase has linear dependence on frequency. Similar behavior for ultrasonic absorption has been reported for viscous liquids as well as various polymers, and this type of absorption is referred to as hysteresis absorption by Hartmann and Jarzynski [16], who postulated that the mechanism responsible for such hysteresis absorption is the trapping of the polymer chain molecules or their side groups in one of many local metastable potential-energy minima. In this case of VDF/TrFE copolymer, it may be necessary to consider the additional absorption due to the phase transition, which is superimposed on the hysteresis absorption. Actually, the data of absorption at the paraelectric region (90 °C) shows non-linear frequency dependence, and the additional components become more explicit at higher frequency ranges above about 3 MHz. In general, it is reported that the absorption is in proportion to

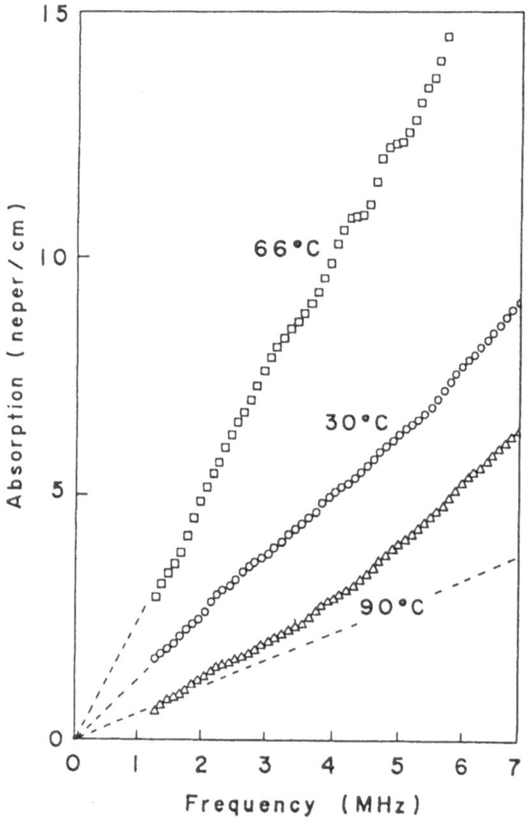

Fig. 19. Frequency dependence of absorption of P(VDF/TrFE) copolymer at three different temperatures

the square of frequency just above the transition point of the second order ferroelectric phase transition. Again, this observation agrees well with the peak temperature lowering phenomenon with increasing frequency described earlier.

On the other hand, Fig. 20 shows the temperature dependence of the velocity at different frequencies of 1, 3, 5, and 7MHz, revealing a critical lowering phenomenon around the Curie temperature, which is characteristic of the ferroelectric phase transition phenomenon. The sound velocity at higher frequencies seems to show this lowering phenomenon at a lower temperature range compared with that at lower frequencies.

As discussed above, the spectroscopic data provide detailed and important information on the phase transition phenomena in polymeric materials, which cannot be obtained by ultrasonic measurements at only a single frequency.

3.4 Relaxation Phenomena at High Temperatures

Polymers undergo various mechanical relaxations with different activation energies. In order to analyze such relaxation processes, it is vital to conduct

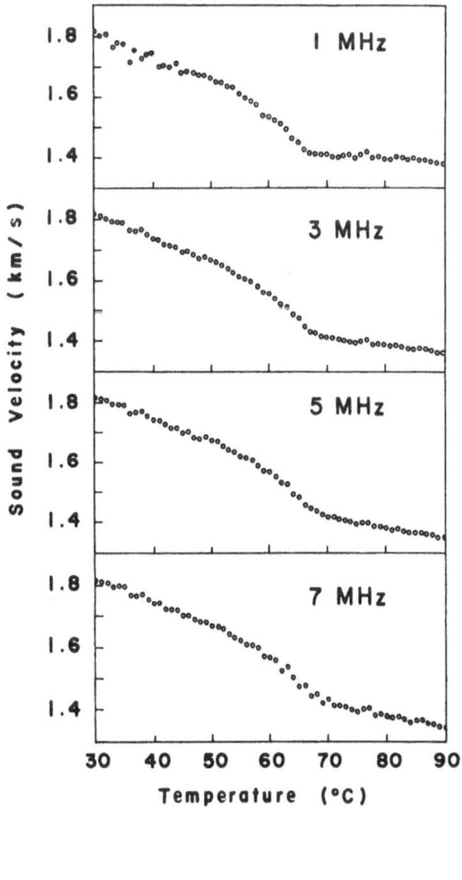

Fig. 20. Temperature dependence of the sound velocity at different frequencies of 1, 3, 5, and 7 MHz around the Curie point of P(VDF/TrFE) copolymer

mechanical relaxation measurements over a wide range of frequencies, and this ultrasonic spectroscopy method was applied to follow the relaxation process of a typical polymer, nylon 6 [17].

Figure 21 shows the high-temperature type measuring cell used. The cell is of a reflection type, and a cooling system for the transducer is added to prevent thermal damage to the polymeric piezo-transducer. The thickness of the nylon 6 sample was 1.97 mm. Since the ultrasonic waves reflect at each interface, several signals appear in time domain, as shown in Fig. 22a, and the path of each signal is shown in Fig. 22b. The first signal (I) corresponds to the one reflected at the bottom of the buffer, the second (II) the one passed through the specimen and reflected at the bottom of the cell, while the third (III) is the one reflected at the bottom of the cell, the upper end of the specimen and again at the bottom of the cell. Therefore, FFT calculation was carried out only for the second signal (II). As shown in Fig. 23, the ratio of a spectral amplitude to that at room temperature decreases, reaches a minimum and then further decreases rapidly as the temperature increases. The temperature where the ratio is a minimum becomes higher as the frequency increases. This minimum indicates the increase of absorption of ultrasonic waves in this temperature range. It is known that nylon 6 has several dispersions as shown in Fig. 24, such as α, β and γ dispersions [18]. The temperatures where the ratios are minima are plotted in the figure by the marks O, which sit on the extension of the reported α dispersion line, implying that the observed minimum is due to the α dispersion. It is known that the α dispersion is related to the motion of the main chain. On the other hand, the rapid decrease of the ratio around 200 °C is due to melting of the specimen, proving that not only the α dispersion but also melting behavior can be observed with this system.

Thus, these results suggest that ultrasonic spectroscopy measurements can be conducted up to at least 200 °C with this high temperature cell using the

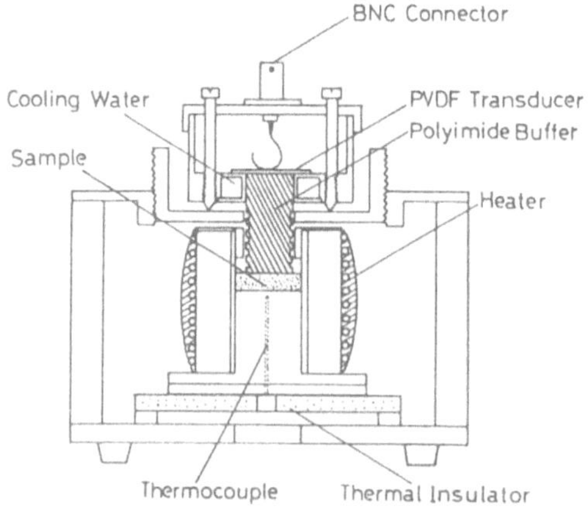

Fig. 21. Details of a high temperature-type measuring cell

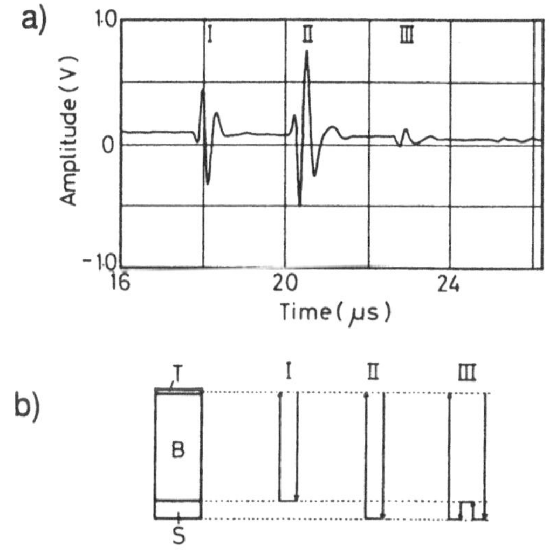

Fig. 22a. Time domain signals from a nylon 6 specimen. **b** Paths of the signals (I), (II) and (III). T, B and S are transducer, buffer and specimen, respectively

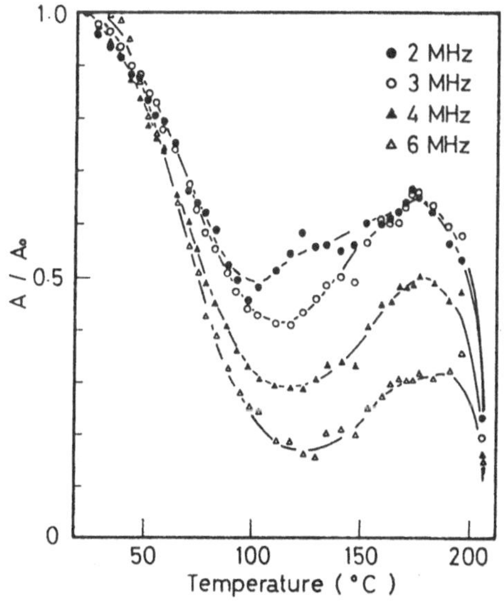

Fig. 23. Temperature dependence of the spectral amplitude of nylon 6 for signal (II) in Fig. 22

polyimide buffer. If one wants to carry out such measurements at even higher temperatures, the following points should be considered; first, the length of the polyimide buffer should be shorter to reduce severe damping of the ultrasonic waves, and, second, a more powerful polymeric transducer, such a VDF/TrFE copolymer film should be explored. Spectroscopic measurements above 300 °C are expected to be used to investigate detailed changes in acoustic properties of liquid crystal polymers or pitches at high temperatures, these being especially important materials for industrial application. Moreover, another subject for

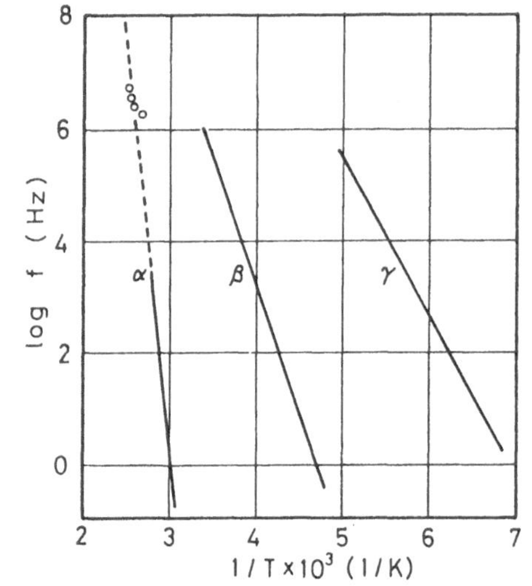

Fig. 24. Dispersion map of nylon 6. O are the data obtained in this ultrasonic spectroscopy study

future study may be widening the measuring frequency range because the polymeric transducer film itself can generate a 500MHz ultrasonic wave [19].

3.5 Detection of Heterogeneities in Uniform Polymer Matrix

Next, the possibility of detecting heterogeneities existing inside a polymeric matrix by the ultrasonic spectroscopic method is discussed [7, 11, 20]. When the test samples are transparent, conventional optical or X-ray methods may be employed for such non-destructive testing. However, these samples are generally opaque and do not give clear contrast for X-rays, and so these conventional methods cannot be applied. Since ultrasonic waves can pass through any material and are very sensitive to boundaries with acoustic mis-matchings, this ultrasonic spectroscopic method may be one of the most suitable methods for identifying not only the existence but also the size of a discontinuity.

To prove the applicability of the ultrasonic spectroscopic method to these non-destructive testings and to clarify further the relationship between the sizes of a discontinuity and the ultrasonic spectra, spectroscopic measurements were carried out on the polymeric (PTFE/glass beads) composite system shown in Fig. 25, which reveals a model compound of uniform dispersing heterogeneities (glass beads) in uniform matrix (PTFE). Figure 26 shows the time domain datum and its power spectrum for the PTFE containing no glass beads, while Fig. 27 shows the results for the composite samples containing 5% of the indicated sizes of glass beads. It is clear that the waveforms and power spectra change drastically with the sizes of included beads. The amplitude for the whole frequency range first becomes lower, and then the amplitude at a lower frequency range recovers gradually as the glass bead sizes become smaller.

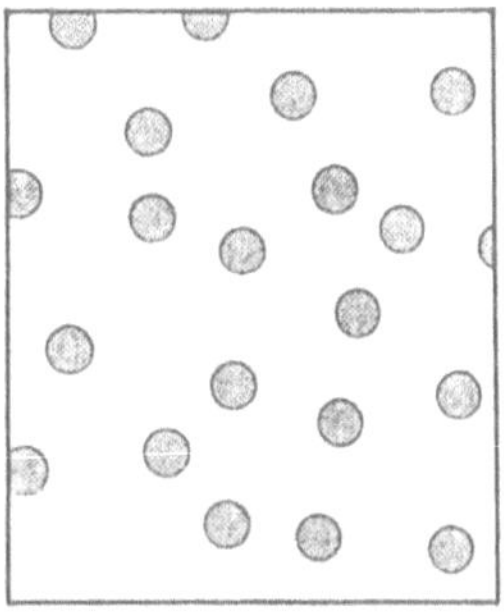

Fig. 25. Schematic illustration of heterogeneities in uniform polymer matrix

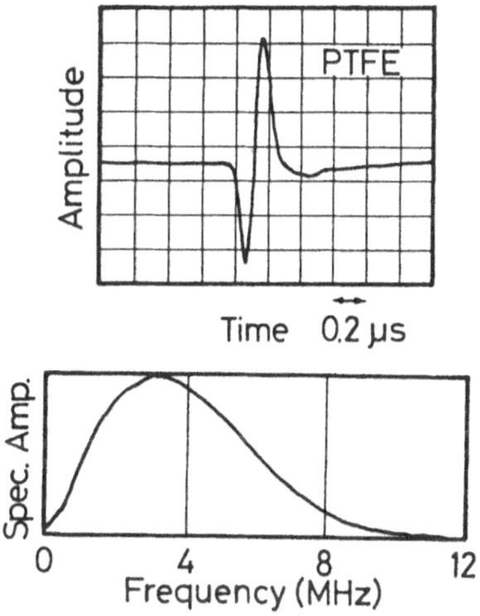

Fig. 26. Time domain signal and power spectra of polytetrafluoroethylene

The scattering of ultrasounds from the included glass beads is thought to play an important roll in such significant spectral changes. In general, the scattering phenomena from composite materials can be classified into the following three categories, depending on the relationship between the wavelength l, and the diameter of the glass bead D. The first category is called Rayleigh scattering, which occurs when l/D is much larger than 1; the second is called stochastic scattering, when l/D is around 1; the third is called diffusion scattering, when l/D is less than 1. In the present case, as the velocity in PTFE is about 1300 m/s, the results for smaller (less than 120 μm) glass bead samples shown in Fig. 27 may be explained by the Rayleigh scattering mechanism, where the attenuation increases proportionally to the fourth power of the frequency and becomes smaller as D becomes smaller. On the other hand, the result for larger (greater than 400 μm) glass bead samples may be explained by the diffusion scattering mechanism, where the attenuation does not have the frequency dependence and becomes larger as D becomes smaller.

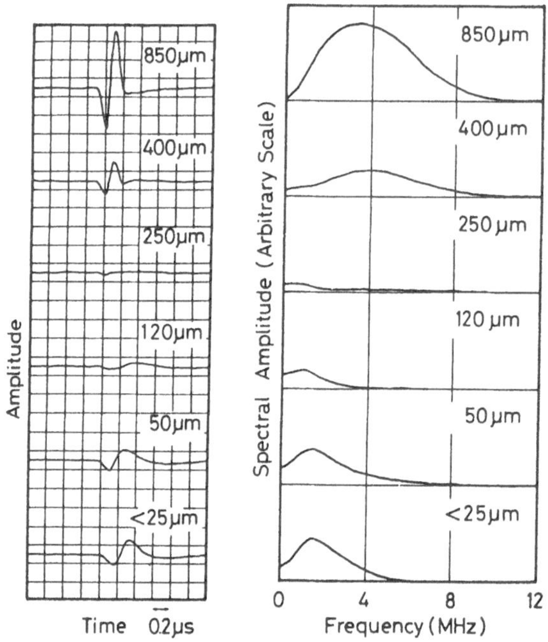

Fig. 27. Time domain signals (*left*) and power spectra (*right*) of polytetrafluoroethylene samples containing 5% of various sizes of glass beads

As mentioned above, the spectral changes observed here can be explained qualitatively. However, if one wants to understand the phenomena quantitatively, other factors such as multiple scattering and the distribution of the glass bead sizes should be considered. The delay of the waveform is probably due to multiple scattering and can be found in the time domain data of Fig. 27 for small (less than 120 μm) glass bead samples, where the glass beads are rather dense and the wavelengths of scattering ultrasound are comparable with or shorter than the distances between the dispersed glass beads. Thus, the contribution of multiple scattering may be confirmed by analyzing further the sound velocity changes in connection with the sound wavelength and the dispersed distances.

Similar data were obtained for 10% bead content Teflon/glass bead samples, as shown in Fig. 28. Again, it is noted that the waveforms and power spectra change drastically with the bead size. The changing behavior is very similar to that for the previous 5% bead content sample, except that the spectral amplitude is smaller. To depict such changing behavior more clearly, the ratios of the spectral amplitude for composite samples to that for pure Teflon are plotted vs frequency in Fig. 29, where the ordinate is the normalized spectral amplitude by pure Teflon data. The data for 5% bead content samples shows a similar result, which consists of two graphs for large bead size samples (850–250 μm) and for smaller bead size samples (less than 250 μm). That is, for the larger bead size

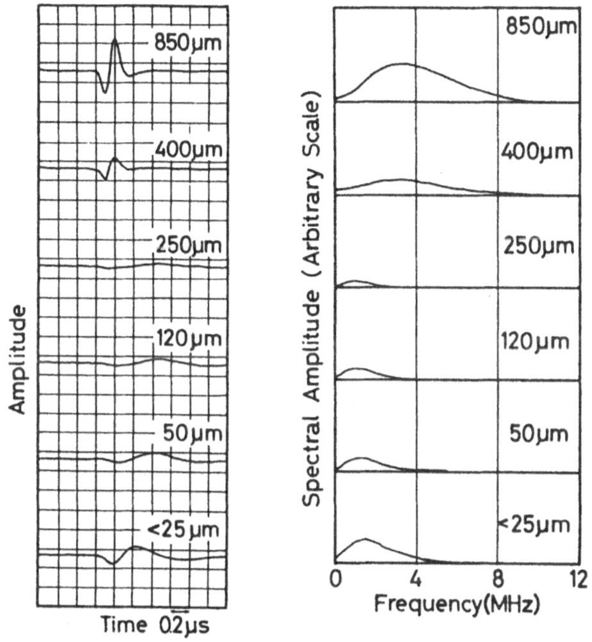

Fig. 28. Time domain waveforms (*left*) and power spectra (*right*) of Teflon composite samples containing 10% of indicated sizes of glass beads

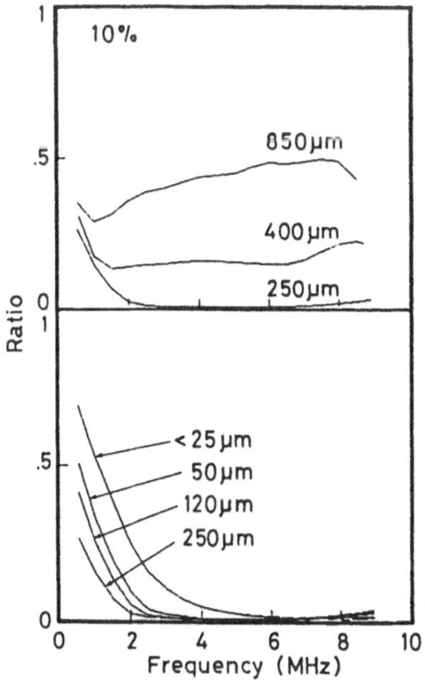

Fig. 29. Frequency dependence of ratios of spectral amplitude for Teflon/glass bead composites (10% bead content) to that for pure Teflon

group, the spectral amplitude decreases almost uniformly for the whole frequency range with decrease in bead size, whereas in the smaller bead size group, the spectral amplitude for the lower frequency range recovers with decrease of bead size.

These results reveal that the scattering of ultrasound at the interface between Teflon matrix and included glass beads plays an important role in the apparent attenuation. Frequency dependence of the scattering cross section of the ultrasounds by a sphere has already been discussed theoretically by Morse and Ingard [21]. They showed that the scattering cross section first increases sharply, and then levels off with increase of $\pi D/l$, where D and l are the diameter of the sphere and wavelength in the matrix, respectively. The initial sharp increase of the scattering cross section corresponds to the Rayleigh scattering mechanism and the leveling-off behavior corresponds to the stochastic and diffusion scattering mechanisms, where the frequency dependence is much weaker than for the Rayleigh scattering. The uniform decrease of spectral amplitude for the large bead size group shown in Fig. 29 is judged to correspond to the leveling-off behaviour of the scattering cross section because the $\pi D/l$ value becomes very large in this group. On the other hand, the curves representing recovery of the spectral amplitude in the low frequency region for small bead size samples are almost identical if the curves are plotted vs $\pi D/l$. This means that the scattering behavior for this bead size region is dominated in the same way.

Since the data shown in Fig. 29 include the three parameters of frequency, bead size, and content, it is impossible to show all parameter dependencies at one time. Plotting data on graphs with different combinations of parameters may reveal interesting aspects. Figure 30 shows the bead size dependence of normalized spectral amplitude for 5% and 10% bead content samples at 2MHz. The bead size at minimum spectral amplitude is about 240 μm which means that $\pi D/l = 1.2$. There is a theory that an attenuation Q^{-1} of elastic wave due to the multiple scattering from the crack [22] reaches a peak value when the l is nearly twice crack length L. Comparing the value of l/D in our study with this value of l/L shows our value to be a little higher. This discrepancy may be ascribable to

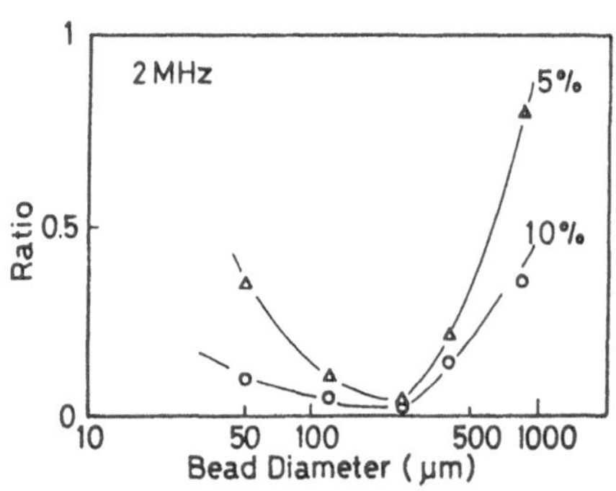

Fig. 30. Bead size dependence of ratios of spectral amplitude at 2 MHz for 5% and 10% Teflon/glass bead composite samples

the difference in the shape of the scatterers or to the effect of multiple scattering. Future clarification of this variance will be interesting.

4 Two-Dimensional and Non-Destructive Analysis

The nondestructive evaluation of CFRP is recognized to be very important and indispensable, because these polymeric composites are utilized for many industrial applications such as planes, spaceships and various load-bearing parts due to their toughness and lightness. Although several methods including X-ray and ultrasonic techniques have already been proposed for the evaluation of defects in CFRP, no adequate method exists, and it is still a difficult task to distinguish between defects and carbon fibers in CFRP. Therefore, the development of new effective methods was necessary.

In this section, the application of a scanning system of single-pulse ultrasonic spectroscopy for the structural evaluation of CFRP (carbon-fiber reinforced plastics) is described. As seen below, the ultrasonic data obtained from a single pulse response in the frequency domain provided higher-contrast images than conventional ultrasonic techniques utilizing a single fixed frequency. Attempts to obtain much higher resolution in images by applying echo analysis developed in earthquake theory [12] and thickness measurements of a thin layer [13, 14] are discussed.

4.1 Two-Dimensional Measuring System

Figure 31 shows the block diagram of the used scanning single-pulse ultrasonic measuring system [3, 23, 24]. Signal processing in this system is carried out as follows. First a short electrical rectangular pulse is generated from a pulse generator (150 V in height H and 25 ns in duration). Subsequently, the pulse is transformed into an ultrasonic pulse by a focus-type transducer. In order to obtain pulse-like ultrasonic waves, the transducer should have a wide-band characteristic, as already mentioned. The ultrasonic waves reflected from the sample were detected with the same transducer, and inversely transformed into an electric signal, which is digitized and recorded with a storage scope (Iwatsu TS-8123: 8 bits, 512 words, fastest sampling frequency is 25 GHz).

Such measurements were repeated for each step of the computer-controlled X-Y stage movements, and later the frequency analysis was performed for every waveform to obtain two-dimensional information for an entire sample and over a wide frequency range. Before the calculation, data were expanded from 512 to 4096 words by adding data of zero value to obtain a higher frequency resolution.

The transducer used in this study had a concave surface (20 mm focal length, 5 mm in diameter) coated with a thin piezoelectric film of vinylidene fluoride

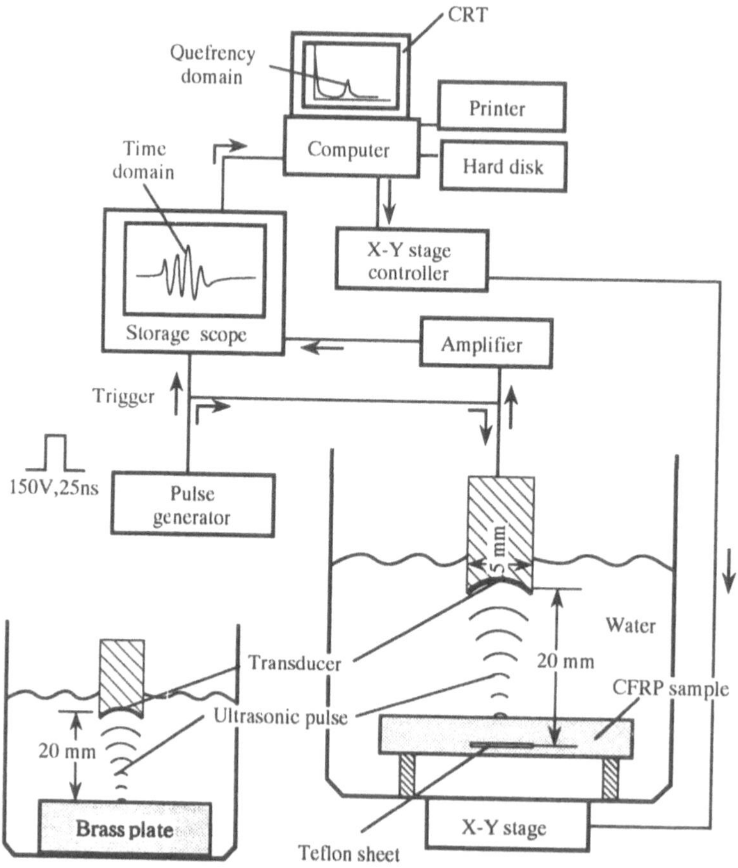

Fig. 31. Schematic diagram of the scanning single-pulse ultrasonic measuring system

and trifluoroethylene copolymer (P(VDF/TrFE)), which has a low acoustic impedance ($\approx 2 \times 10^6 \mathrm{Pa \cdot s/m}$) and wide frequency characteristics (mid-frequency is 25 MHz).

Figure 32 shows the schematic structure of the tested CFRP sample. The CFRP sample had a multilayered structure of CFRP cloth. The carbon fibers in one layer of cloth were aligned unidirectionally and the adjacent layers of cloth were set cross-directionally. The thickness of CFRP cloth was 70 μm, and the examined CFRP sample was composed of 72 layers of cloth (about 5 mm thickness in all). In order to model the separation of the CFRP layers (defect), a Teflon sheet (80 μm in thickness) was sandwiched between the 54th and the 55th layers of CFRP cloth, at a level 3.8 mm below the CFRP top surface. The sound velocity and the acoustic impedance of the CFRP sample are about 3×10^3 m/s and 5×10^6 Pa·s/m, respectively. Water was used as the coupling medium because of its low ultrasonic attenuation characteristics, and the CFRP

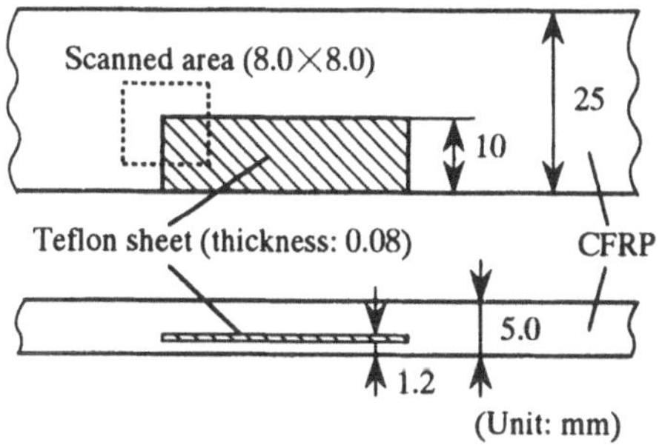

Fig. 32. Schematic views of the CFRP sample including the Teflon sheet, which models the separation of the CFRP layers

sample was held in a water bath as shown in Fig. 31. All measurements were carried out at room temperature.

4.2 *Analysis of Defects in CFRP (Carbon Fiber Reinforced Plastics)*

4.2.1 *Characteristics of Measuring System*

To obtain the term $(1 + \gamma^2 + 2\gamma \cos 2\pi ft)$ from $|A(f)|^2/|X(f)|^2$, the values of $|A(f)|$ and $|X(f)|$ are necessary as already shown in Eq. (28). However, we use $|G(f)|^4 |E(f)|^2$ instead of $|X(f)|^2$ as shown in Eq. (29), since $|X(f)|$ cannot be obtained experimentally:

$$
\begin{aligned}
|A(f)|^2 &= |X(f)|^2(1 + \gamma^2 + 2\gamma \cos 2\pi f\tau) \\
&= |G(f)|^4 \cdot |E(f)|^2 \\
&\quad \times \{|H_1(f)|^2 \cdot |K(f)|^2 \cdot |H_2(f)|^2(1 + \gamma^2 + 2\gamma \cos 2\pi f\tau)\}
\end{aligned}
\tag{29}
$$

where $|G(f)|^4|E(f)|^2$ was measured experimentally with a brass plate instead of the CFRP sample as shown in Fig. 31. In this case, we may assume that $|H_1(f)| \approx |H_2(f)| \approx |K(f)| \approx 1$, because the ultrasonic attenuation in water is very small and the acoustic impedance of the brass plate ($\approx 4 \times 10^7$ Pa·s/m) is very different from that of water ($\approx 1.5 \times 10^6$ Pa·s/m).

In this system, since the resolution of the analog-digital (A/D) converter is about 1% (8 bits), the measurable range of the power spectrum is limited to about four figures. Therefore, for precise analysis, the square of the transfer function of the measuring system must be flat within two or three figures. Figure 33 shows the waveform reflected on a brass plate (Fig. 33(a)) and its power spectrum (Fig. 33(b)), revealing the maximum peak at about 18 MHz.

4.2.2 *Original Waveform of CFRP Sample*

Figure 34a shows schematic views of the CFRP sample, ultrasonic signals and the position of the transducer, while Fig. 34b shows the observed waveforms

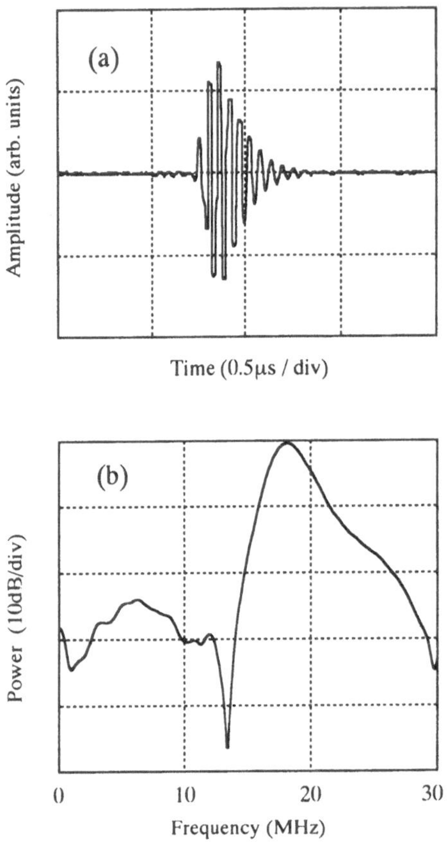

Fig. 33a. Ultrasonic waveform of reflection from the brass plate. **b** Power spectrum of this reflection

reflected at different points A, B and C. The waveforms were analyzed numerically to obtain their power spectra and the autocorrelation function, as shown in Fig. 35a, b respectively. Here, we first treat the original waveforms and the power spectra; correction using Eq. (29) will be discussed in the next section.

At measuring point A where there is no Teflon sheet, the time-domain waveform reveals only one reflection from the bottom surface of the CFRP sample, and its power spectrum shows a decrease in the higher-frequency component above 20 MHz, as shown in Fig. 35a. The decrease is probably caused by ultrasonic scattering by the multilayered structure of CFRP cloth, since the half-wavelength at 20 MHz is about 75 µm, which is comparable to the thickness of CFRP cloth (70 µm).

Next, at measuring point B where the edge of the Teflon sheet is located, the time-domain waveform shows only reflection from the Teflon sheet. The disappearance of the reflection from the bottom surface of the CFRP sample is assumed to be due to the existence of a low-density region in the neighborhood of the Teflon sheet edge, which reflects almost all the ultrasonic waves because of its lower acoustic impedance.

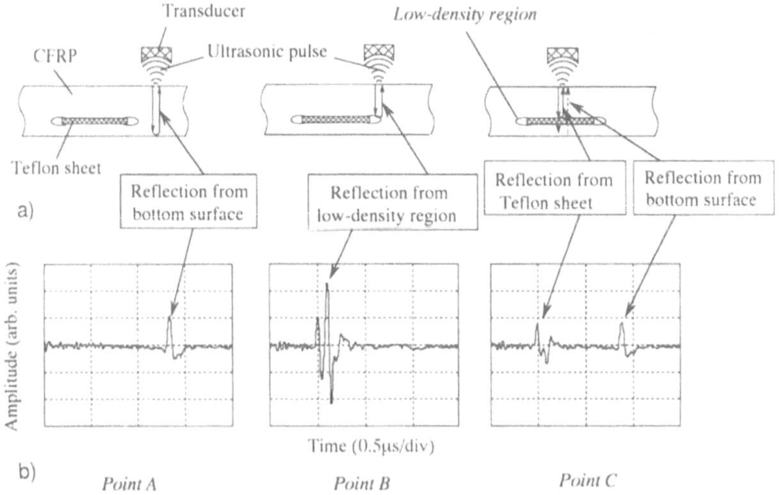

Fig. 34a. Schematic views of ultrasonic reflections at three different points of CFRP specimen. **b** Corresponding time-domain waveforms of reflections

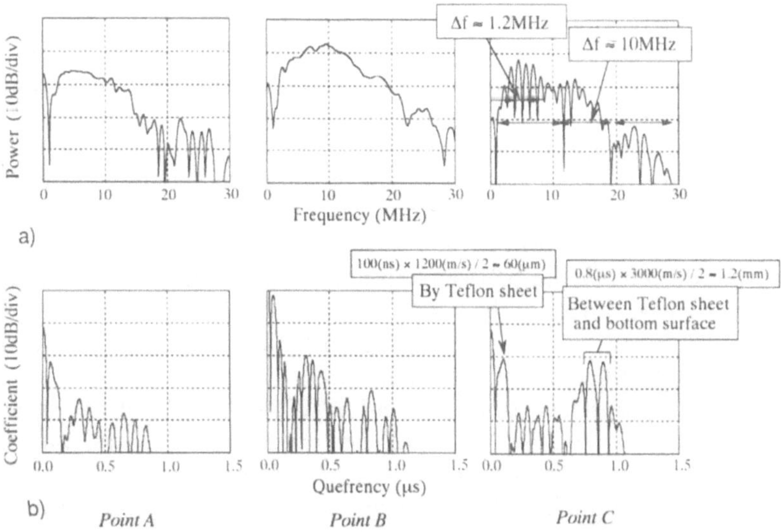

Fig. 35a. Power spectra of original waveforms at three measuring points of CFRP specimen. **b** Corresponding autocorrelation coefficients

On the other hand, at measuring point C above the Teflon sheet, the reflection from the bottom surface of the CFRP sample becomes as large as that from the Teflon sheet. The reflection from the Teflon sheet may be divided into two simple reflections from the upper and lower surfaces. The power spectra at point C shows periodic minima with 1.2 MHz frequency intervals, and the autocorrelation function shows a peak at about 0.8 μs, as shown in Fig. 35. Such periodic power spectra are thought to originate from interference of the waves reflected at different positions. The autocorrelation function reveals peaks corresponding to the periodic power spectra, as mentioned in Sect. 2. When two waves are reflected at different points with separation L and transmitted in a medium with sound velocity of V, the power spectrum for the superposed waves reveals the periodic minima of Δf, as a result of the interference of these two waves, as expressed by [3, 14]

$$L = \frac{V}{2\Delta f}. \tag{30}$$

We can rewrite Eq. (30) using the position of the peak in the autocorrelation function (T_p in frequency domain; $T_P = 1/\Delta f$):

$$L = \frac{VT_p}{2}. \tag{31}$$

In the case of the CFRP sample, the sound velocity V is about 3×10^3 m/s. Equation (31) gives the value of the distance L as 1.25 mm, which is close to the value of the distance between the Teflon sheet and the bottom surface of the CFRP sample tested here (1.2 mm). The power spectrum at this point C exhibits 10 MHz frequency intervals Δf, which correspond to a peak at about 100 ns in the autocorrelation function in addition to the peak at about 0.8 μs. If the value of 1.2×10^3 m/s is employed for the sound velocity in the Teflon sheet, the peak at 100 ns results in the value of 60 μm for the theoretical thickness of the Teflon sheet; this roughly agrees with the 80 μm thickness of the actual Teflon sheet. The difference of 20 μm is probably due to the increase in the sound velocity in the Teflon sheet or a decrease in the thickness of the actual Teflon sheet due to compressional force upon composite formation.

The two peaks at about 0.8 μs in the autocorrelation function are thought to originate from the interference between upper and lower surfaces of the Teflon sheet and the bottom surface of the CFRP sample. Therefore, the difference of position between the two peaks agrees with the position of the peak at 100 ns corresponding to the thickness of the Teflon sheet.

Incidentally, the cepstrum (Fourier transformation of a logarithmic spectrum) is used commonly in seismology [12] and acoustics in the same way as autocorrelation. The merit of taking logarithms of the spectrum is as follows. If a power spectrum of a multiple echo signals (scattering) is expressed as the following equation, Eq. (32), the product in Eq. (32) becomes the sum of the logarithms of each term as in Eq. (33):

$$|A(f)|^2 = |Y(f)|^2(1 + \gamma_2^2 + 2\gamma_2 \cos 2\pi f\tau_2)(1 + \gamma_1^2 + 2\gamma_1 \cos 2\pi f\tau_1) \tag{32}$$

$$\log(|A(f)|^2) = \log(|Y(f)|^2) + \log(1 + \gamma_2^2 + 2\gamma_2 \cos 2\pi f\tau_2) + \log(1 + \gamma_1^2 + 2\gamma_1 \cos 2\pi f\tau_1) . \quad (33)$$

When $\gamma_1, \gamma_2 \ll 1$, the same situation as for normal seismic waves, Eq. (33) is approximated by Eq. (34):

$$\log(|A(f)|^2) = \log(|Y(f)|^2) + 2\gamma_2 \cos 2\pi f\tau_2 + 2\gamma_1 \cos 2\pi f\tau_1 . \quad (34)$$

Then, the cepstrum function (Fourier transformation of Eq. (34)) has peaks corresponding to τ_1 and τ_2.

In this case, however, the cepstrum function at point C as shown in Fig. 36 is spiky and there are no notable characteristics. This may be due to the relationship $\gamma \approx 1$ in this condition, wherein the reflection occurred at very close points.

As described above, the analyzed autocorrelation function reveals quite clearly the details of the internal structure of the CFRP sample, such as the thickness and position of inserted Teflon sheet. Therefore, if we want to evaluate structures with a certain diameter, separation and thickness, we should select the appropriate frequencies. For example, the Teflon sheet produces a peak at about 0.8 μs in the frequency domain. Then, 0.8 μs should be selected to mark the position of the Teflon sheet. Figure 37a shows the two-dimensional distribution of the autocorrelation coefficient of 0.8 μs (8 × 8 mm area with 0.5 mm step), where the Teflon sheet and the low-density region are imaged with highest contrast. Such a clear and position-sensitive image has never been obtained by conventional ultrasonic techniques. However, if an inappropriate coefficient such as 0.6 μs is selected, the two-dimensional distribution shows only the flat image shown in Fig. 37b.

4.2.3 Relative Power Spectra

In this section, we deal with the correction by Eq. (29). Figure 38 shows the relative power spectra obtained by dividing the original power spectra by the power spectrum of the data taken without the sample. It is noted that the power spectra show flat characteristics below 14 MHz and decrease more quickly in the higher-frequency region above 15 MHz than those in Fig. 37. This frequency

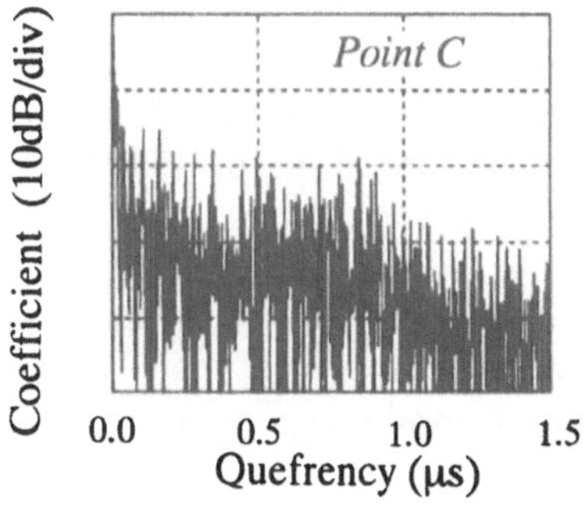

Fig. 36. Cepstrum coefficient of original waveforms at measuring point C

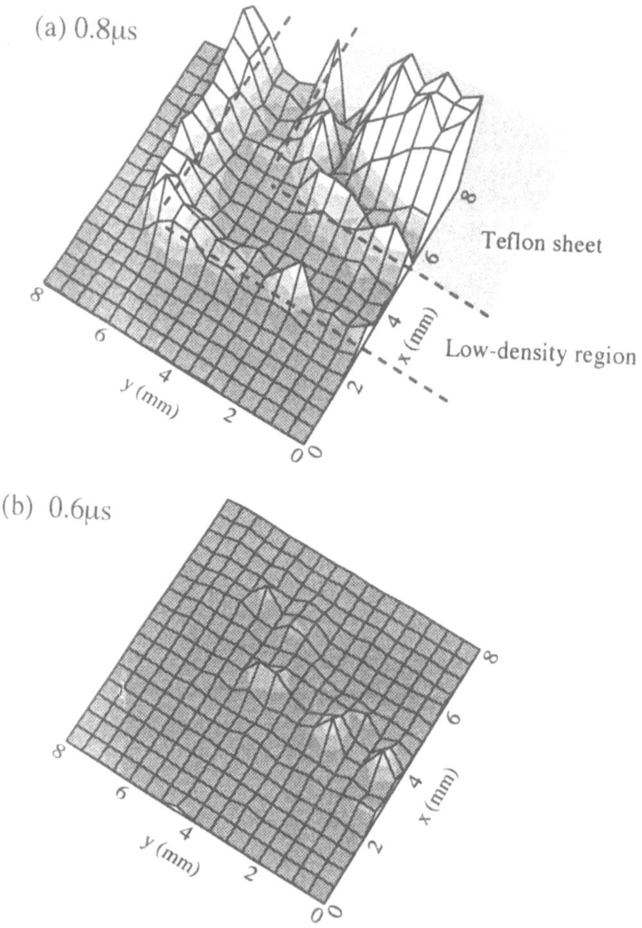

Fig. 37a, b. Two-dimensional autocorrelation images of CFRP including the Teflon sheet with the autocorrelation coefficients: (**a**) of 0.8 μs; (**b**) of 0.6 μs

characteristic suggests that the ultrasonic scattering by the multilayered structure of CFRP cloth occurs to a great extent for wavelengths less than the thickness of CFRP cloth. Therefore, to obtain a flat frequency characteristic in these measurements, it is desirable to use an ultrasonic pulse with an extremely intense amplitude above 15 MHz. Furthermore, with this correction, the peaks of the autocorrelation function at about 0.8 ms became much sharper than those in the original (Fig. 37b).

For correction by Eq. (29), we obtained the relative power spectra by dividing the original power spectra by that at position B, as shown in Fig. 39. It is obvious that the frequency intervals in Fig. 39a are more evident than those in Fig. 35b or Fig. 38b. However, the autocorrelation peaks in Fig. 37b are less

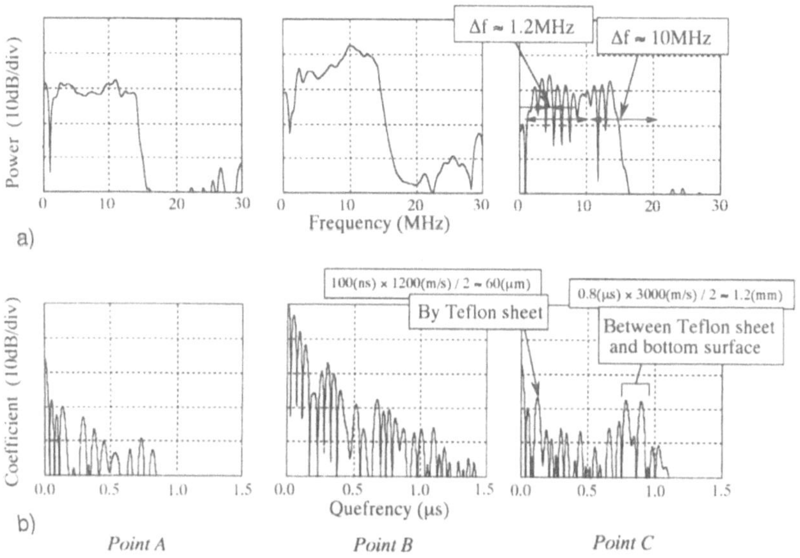

Fig. 38a. Relative power spectra (divided by power spectrum of ultrasonic pulse) at three measuring points. **b** Corresponding autocorrelation coefficients

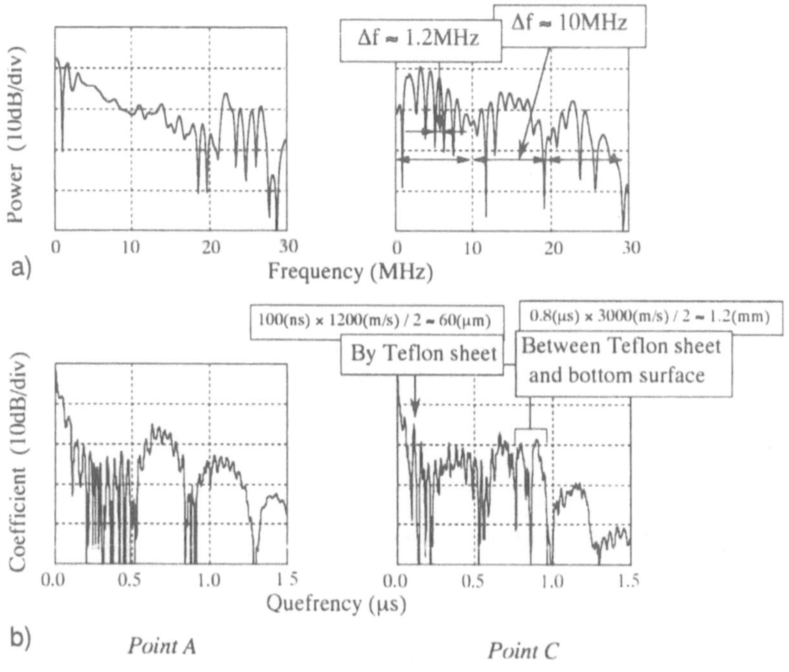

Fig. 39a. Relative power spectra (divided by power spectrum at measuring point B). **b** Their autocorrelation coefficients at two measuring points A and C

clear. The reason for this is probably insufficient digits of the 8-bit A/D converter; if we use sufficient digits with the A/D converter in the future, we will obtain a clearer cosine curve in the frequency domain and sharper autocorrelation peaks in the frequency domain.

As described above, an ultrasonic single-pulse response in the frequency domain was utilized to image the defects in a technologically important polymeric composite of CFRP. This imaging technique provides useful information about the internal structure of the CFRP sample, and is likely to become a powerful method for the nondestructive inspection of many polymeric composites.

5 References

1. Matsushige K (1991). In: Baer E, Moet A (eds) High performance polymers. Hanser, Munich Vienna New York, p 103
2. Matsushige K, Takemura T (1984). In: McGonnagle WJ (ed) International advances in nondestructive testing, vol 10. Gordon and Breach Science, New York London Paris Motreux Tokyo, p 61
3. Takeshita S, Taki S, Matsushige K (1990) Mater Eval 48: 1473
4. Kräutkramer J (1959) Arch Eisenheutten W 30: 693
5. Gericke OR (1963) J Acoust Soc Amer 35: 364
6. Gericke OR (1965) Mater Res Stand 5: 23
7. Whalley HL, Cook KV (1970) Mat Eval 28: 61
8. Brown AF (1982) Ultrasonic testing. John Wiley & Sons, N Y, chap 5
9. Matsushige K, Okabe H, Shichijyo S, Takemura T (1985) Jpn J Appl Phys 24–1: 34
10. Murayama N, Obara H (1993) Jpn J Appl Phys 22–3: 3
11. Hiramatsu N, Taki S, Matsushige K (1988) 27–1: 26
12. Bogert BP, Healy MJR, Tukey JW (1963). In: Rosenblatt M (ed) Time series analysis. Wiley New York, p 209
13. Delebarre C, Rouvaen JM, Bruneel C, Frohly J (1988) J Appl Phys 63: 1846
14. Houze M, Nongaillard B, Gazalet M, Rouvaen JM, Bruneel C (1984) J Appl Phys 55: 194
15. Okabe H, Taki S, Shichijyo S, Matsushige K, Takemura T (1986) Jpn J Appl Phys 25–1: 64
16. Hartmann B, Jarzynski J (1972) J Appl Phys 43: 4304
17. Hiramatsu N, Taki S, Matsushige K (1987) Jpn J Appl Phys 26–1: 76
18. Wada Y (1964) Koubunshi no butsuri (Polymer physics) Asakura Shoten, Tokyo, p 166 [in Japanese]
19. Sussuner H, Michas D, Asstalg A, Hunklinger S, Dransfeld K (1973) Phys Lett 45A: 475
20. Hiramatsu N, Taki S, Matsushige K (1987) Jpn J Appl Phys 28–1: 33
21. Morse PM, Ingard KU (1968) Theoretical acoustics. McGraw-Hill Publ New York
22. Kikuchi M (1981) Phys Earth Planetary Interior 25: 159
23. Okabe H, Taki S, Matsushige K (1992) Jpn J Appl Phys 31: 96
24. Okabe H, Matsushige K (1992) : Jpn J Appl Phys 32 Part 1: 3621

Editor: Prof. Sir Edwards
Received November 1994

Author Index Volumes 101-125

Author Index Vols. 1-100 see Vol. 100

Subject Index

Errata

Unfortunately, typographical errors occured in Vol. 114 of this series, pp. 283. Please find here the correct equations:

$$g^{(1)}{}_{Hv;h}(\tau;\, q_\perp, q_{||}) = \exp[-\tau/\hat{\tau}_T(q_\perp, q_{||})] \qquad \text{Case 2} \qquad (113)$$

$$g^{(1)}{}_{Hv;h}(\tau;\, 0, q_{||}) = \exp[-\tau/\tau_B] \qquad \text{Case 2a} \qquad (114)$$

$$g^{(1)}{}_{Hv;h}(\tau;\, q_\perp, 0) = \exp[-\tau/\tau_T] \qquad \text{Case 2b} \qquad (115)$$

$$g^{(1)}{}_{Hh;h}(\tau;\, q_\perp, q_{||}) = \exp[-\tau/\hat{\tau}_S(q_\perp, q_{||})] \qquad \text{Case 3} \qquad (116)$$

Springer-Verlag and the Environment

We at Springer-Verlag firmly believe that an international science publisher has a special obligation to the environment, and our corporate policies consistently reflect this conviction.

We also expect our business partners – paper mills, printers, packaging manufacturers, etc. – to commit themselves to using environmentally friendly materials and production processes.

The paper in this book is made from low- or no-chlorine pulp and is acid free, in conformance with international standards for paper permanency.